POLYMER SCIENCE AND TECHNOLOGY

POLYMER-MATRIX COMPOSITES

MATERIALS, MECHANICS AND APPLICATIONS

POLYMER SCIENCE AND TECHNOLOGY

Additional books in this series can be found on Nova's website under the Series tab.

Additional e-books in this series can be found on Nova's website under the eBooks tab.

POLYMER SCIENCE AND TECHNOLOGY

POLYMER-MATRIX COMPOSITES

MATERIALS, MECHANICS AND APPLICATIONS

ELIAS RICE
AND
BETH SPARKS
EDITORS

nova
science publishers
New York

NOTICE TO THE READER

Library of Congress Cataloging-in-Publication Data

ISBN: 978-1-53613-398-1

Published by Nova Science Publishers, Inc. † New York

CONTENTS

PREFACE

This books opens with an evaluation of the effects of the acid environment on sulfur-polymer matrix composite, with Portland cement-based composite chosen as a comparative material. Unlike conventional Portland cement-based composites, sulfur-polymer matrix composites are produced without water and achieve final strength in few days. Following this, the effects of electron beam irradiation on copper oxide added polymer matrix composite was investigated. Copper oxide added polymer matrix composite has been known for engineering applications that involving plastics moving parts to reduce friction. In a separate study, inorganic Montmorillonite nanofibers and organic cellulose nanofibers obtained through acid hydrolysis were simultaneously added to plasticized starch with defined CNF-to-MMT ratio and processed into nanocomposites by melt extrusion. The continuously isothermal aerobic ageing on samples was carried out by oven-heating at 60°C so as to investigate the organic nanofillers as an alternative to inorganic nanofillers. The authors also investigate the physico-mechanical characteristic of low density polyethylene (LDPE) and starch blends when undergone acid and alkaline attacks. In the following chapter, simulation of the temperature distribution in textile composite structures is achieved by utilizing an analytical-symbolic approach. Analytical heating solutions along each yarn provide accurate solutions with far fewer nodes than numerical solutions, and require the minimum number of symbolic network nodes. The final investigation was conducted to investigate the effects of copper (II) oxide (CuO) composition on the degradability of high density polyethylene/ethylene vinyl acetate (HDPE/EVA) blends. It was observed that both degree of crystallinity and crystallites sizes of the composited increased after the ageing operation.

Chapter 1 - Polymers had a great influence in the development of practical applications of materials and among the extensive variety of these compounds, organic polymers had a crucial role. Having in mind their wide practical application and equally significant function in divers technological processes, inorganic polymers were unduly neglected. In combination with some other materials, inorganic polymers make high-performance composite materials with wide and sustainable use. A huge number of inorganic materials is capable of making polymer chains and beyond them, sulfur deserves a particular place.

Sulfur- polymer matrix composites are high performance environmentally sustainable and durable thermoplastic materials made of mineral aggregates, filler and modified sulfur binder which replaces cement and water in conventional Portland cement composite at temperatures above the hardening point of sulfur (120°C). Different types of modified sulfur have been created in order to prevent failure of composite material with elemental sulfur as a binder. For this purpose, various chemical modifiers were used to polymerize sulfur and one of them is dicyclopentadiene which reacts with elemental sulfur to form long-chain polymeric polysulfides. Unlike conventional Portland cement- based composites, sulfur- polymer matrix composites are produced without water and achieve final strength in a few days. In spite of durability and sustainability of sulfur- polymer matrix composites, their wide use is limited due to the high price of chemical modifiers.

This research evaluates the effects of the acid environment on sulfur-polymer matrix composite and Portland cement- based composite is going to be used as a comparative material. For this purpose morphological analysis of the samples exposed to acid influence was done. Based on the obtained results, it is evident that the investigated materials had different responses to the acid exposure. Unlike Portland cement- based composite, insignificant changes in the pore system of sulfur- polymer matrix composite in terms of deterioration, pore distribution and homogeneity were observed. Namely, no signs of severe damage were detected at the surface and inside the sulfur- polymer matrix composite samples. The superior properties of sulfur- polymer matrix composite over those of Portland cement- based composite, recommend that sulfur- polymer matrix composite can be effectively used as a construction material in acid environment.

Chapter 2 - In this work, the effects of electron beam irradiation (EB) on copper (I) oxide added polymer matrix composite (PMC) was investigated. Low density polyethylene (LDPE) was blended with copper (I) oxide using an extruder while further undergone EB at 50-300 kGy to induce crosslinking in the PMC. Such copper (I) oxide added PMC has been known for engineering applications that involving plastics moving parts to reduce friction. It was found that the EB has interacted with copper (I) oxide to form crosslinking network in the PMC matrix for stronger structures as exhibited by the gel content of PMC were higher than pristine LDPE at higher irradiation dosages (≥100 kGy). The formation of crosslinking networks in polymer matrix of PMC could resist the structural reorganization of polymer matrix when subjected to straining stress. The occurrence of fibrils on the facture surface of the PMC proved that EB crosslinking enables evenly load transfer when subjected to straining as well. Finally, the application of irradiation dosage up to 100 kGy and 200 kGy also significantly increased the crystallinity of PMC by the formation of crosslinking networks. The formation of crosslinking networks in LDPE matrix could rearrange the random chains structures into highly ordered chains arrangement. This is due to the EB irradiation could induce the formation of new crystallites in amorphous regions of the PMC.

Chapter 3 - The inorganic Montmorillonite (MMT) and organic cellulose nanofibers (CNF) obtained through acid hydrolysis were simultaneously added to plasticized starch with a defined CNF-to-MMT ratio and processed into nanocomposites by melt extrusion. In order to investigate the possibility of organic nanofillers an alternative to inorganic nanofillers in sustaining the composite properties for food container application under thermal exposures, the continuously isothermal aerobic ageing on samples was carried out by oven-heating at 60°C. Measurements of tensile properties, impact strength, water absorption and cross-sectional morphology were conducted on TPS and nanocomposites as a function of aging time (0-, 6-, 12-, 18-, 24-, 48- and 72-hours). The changes of internal structure or performance of studied materials varied with the different ageing stages. The tensile results elucidated the increase of material compactness by formation of crosslinking during initial ageing to decrease the chain mobility. As compared to MMT, CNF is more suitable to serve as a post cure agent since MMT has inhibited the increment of TS. The strength/modulus are found to be decreased during middle ageing stage which related to the degradation induced by intensive thermal exposure, but the ensuing molecular rearrangement of scissioned remainders tend to recover properties. Changes in morphology of fractured samples after thermal aging revealed the modification of TPS morphology fracture surface by crosslinking network. The obtained degree of crosslinking density could be reflected by % water absorption. The impact loading response manifested the high brittleness of thermal-aged-samples based on unfavourably deteriorated impact resistance.

Chapter 4 - This research investigated the effects of different loading level ratios of $CuO/MnCl_2$ and electron beam irradiation on the properties of thermoplastic starch (TPS). The increasing of manganese (II) chloride loading level has gradually improved the tensile strength and Young's modulus of all un-irradiated and irradiated $CuO/MnCl_2$ added TPS. This is because the manganese (II) chloride could form a strong interaction with starch molecules in TPS matrix. Besides, the increasing of copper (II) oxide loading level has significantly improved the thermal stability of TPS matrix by delaying the thermal degradation temperatures of TPS matrix. The copper (II) oxide in TPS matrix could enhance the thermal stability of TPS by acting as inorganic retarder during combustion process. The application of electron beam irradiation could interact well with manganese (II) chloride particles in TPS matrix by inducing the crystallinity of TPS matrix. This interaction effect could increase the crystallinity of TPS by converging the random polymer chains arrangement in amorphous region of TPS matrix into more ordered structure.

Chapter 5 - This chapter aims at investigating the physico-mechanical characteristic of low density polyethylene (LDPE) and starch blends when undergone acid and alkaline attacks. It was found that the mechanical properties of the polymer blends have improved at the initial stage. This is because the acid solution (hydrochloric acid) and the alkaline solution (sodium hydroxide) consumed the starch portion enables the remaining LDPE chains to rearrange in highly consistent structure without the disruption of starch molecules. However, as the duration of exposure increased, the polymer blends undergone degradation and thus resulted in poor tensile properties. It was also found that the X-ray diffraction peaks become broader at initial treatment stage indicating formation of crystallinity structure. Under prolong exposure to acid and alkaline conditions whereby starch has been fully consumed, the acid and alkaline components attacked the LDPE bonds directly. This can be evidenced from the electron microscope observation of polymer blends after long period of exposure where the matrix showed discontinuity along with voids with short and thick fibrils.

Chapter 6 - In this chapter, simulation of the temperature distribution in textile composite structures is achieved by utilizing an analytical-symbolic approach. Analytical heating solutions along each yarn provide accurate solutions with far fewer nodes than numerical solutions, and require the minimum number of symbolic network nodes. The lower upper (LU) decomposition of the symbolic network scales as the cube of the number of nodes. Multiple evaluations, including iterating temperature-dependent thermal conductivity to achieve a self-consistent solution, scaled linearly with the number of nodes and hardly affected the total solution time. Results show that the memory consumption, CPU time, and solutions of the proposed network calculation method, compare favorably to a finite element analysis.

Chapter 7 - This study was conducted to investigate the effects of copper (II) oxide (CuO) composition on the degradability of high density polyethylene/ethylene vinyl acetate (HDPE/EVA) blends. The content of

CuO incorporated was varied from 0 – 5 phr and the compounded composites were subjected to ramp-120°C and ramp-100°C degradation. The tensile results showed that the all composites showed deterioration of mechanical properties after the degradation process. In terms of thermal properties, the differential scanning calorimatry (DSC) analysis suggested that both the melting temperature (T_m) and melting heat (ΔH_m) increased after 6 days degradation and the values were observed to reduce again when the degradation prolonged. Similar trend was observed to the temperature at 5% weight loss ($T_{5\%}$) of composites for the thermogravimetric analysis (TGA). The physical properties of the composites have been investigated through the X-ray diffraction (XRD) analysis. It was observed that both degree of crystallinity and crystallites sizes of the composited increased after the ageing operation. Finally, the impacts of the CuO composition on the degradability of the composites have been verified through the SEM analysis.

In: Polymer-Matrix Composites ISBN: 978-1-53613-398-1
Editors: Elias Rice and Beth Sparks

Chapter 1

The Synthesis of a Sulfur-Polymer Matrix Composite and Morphological Analysis of the Samples in Extreme Conditions

Milica M. Vlahović*[*] *and Sanja P. Martinović
Institute of Chemistry, Technology and Metallurgy,
University of Belgrade, Belgrade, Serbia

Abstract

Polymers had a great influence in the development of practical applications of materials and among the extensive variety of these compounds, organic polymers had a crucial role. Having in mind their wide practical application and equally significant function in divers technological processes, inorganic polymers were unduly neglected. In combination with some other materials, inorganic polymers make high-performance composite materials with wide and sustainable use. A huge number of inorganic materials is capable of making polymer chains and beyond them, sulfur deserves a particular place.

[*] Corresponding Author Email: milica.vlahovic@yahoo.com.

Sulfur- polymer matrix composites are high performance environmentally sustainable and durable thermoplastic materials made of mineral aggregates, filler and modified sulfur binder which replaces cement and water in conventional Portland cement composite at temperatures above the hardening point of sulfur (120°C). Different types of modified sulfur have been created in order to prevent failure of composite material with elemental sulfur as a binder. For this purpose, various chemical modifiers were used to polymerize sulfur and one of them is dicyclopentadiene which reacts with elemental sulfur to form long-chain polymeric polysulfides. Unlike conventional Portland cement-based composites, sulfur- polymer matrix composites are produced without water and achieve final strength in a few days. In spite of durability and sustainability of sulfur- polymer matrix composites, their wide use is limited due to the high price of chemical modifiers.

This research evaluates the effects of the acid environment on sulfur-polymer matrix composite and Portland cement- based composite is going to be used as a comparative material. For this purpose morphological analysis of the samples exposed to acid influence was done. Based on the obtained results, it is evident that the investigated materials had different responses to the acid exposure. Unlike Portland cement- based composite, insignificant changes in the pore system of sulfur- polymer matrix composite in terms of deterioration, pore distribution and homogeneity were observed. Namely, no signs of severe damage were detected at the surface and inside the sulfur- polymer matrix composite samples. The superior properties of sulfur- polymer matrix composite over those of Portland cement- based composite, recommend that sulfur- polymer matrix composite can be effectively used as a construction material in acid environment.

Keywords: sulfur-polymer matrix composite, Portland cement composite, morphological analysis, homogeneity

1. Introduction

A composite is a material containing different and distinct phases: matrix and one or more dispersed phases. Its bulk properties significantly differ from those of any of the constituents.

Matrix represents the primary, continuous phase that holds the dispersed phase and shares a load with it. It is usually less hard phase and more ductile.

Dispersed phase is the secondary, discontinuous phase or phases, embedded in the matrix. It is usually stronger phase and therefore sometimes it is called reinforcing phase.

Various materials contain dispersed phases, but are not composites because of the properties similar to those of their constituents.

The classification of composite materials based on the nature of the matrix:

- Metal matrix composites,
- Ceramic matrix composites, and
- Polymer matrix composites.

The polymer matrix composites are not as strong and heat- resistant as metal matrix and ceramic matrix composites.

The polymeric materials based on inorganic constituents are very important in nowadays practice, therefore, from the aspect of the modern technologies and process solutions, these polymers can play a significant role in various technological processes. The difficulty is that their production on an industrial scale is still problematic. However, inorganic polymers, in combination with other materials, can make composite materials of very high quality and possibilities of application that can in the current situation find extremely sustainable use. Plenty of inorganic materials are capable to generate polymer chains in some circumstances, and beside them special attention should be paid to sulfur.

Sulfur is an element that was being removed from the atmosphere by slow natural processes during the eons. In one of these natural processes of sulfur removal, sulfur was binding to metals giving ores, while in the other natural cycle, sulfur was binding to organic materials through food chains, which led to its storage in crude oil. Thanks to the modern development of industrial society during the last two centuries, the technological revolution, extremely huge amounts of sulfur as a by-product primarily from exploitation of minerals and crude oil have been obtained. Thus sulfur has returned to the environment by various activities representing the reverse processes to the natural process of removing sulfur from the atmosphere [1, 2].

Sulfur can be found in the earth's crust, in the ocean and even in meteorites and it is the 16th most abundant element in nature. Sulfur occurs naturally all over the world and it can be very prolific on the places where sulfur-rich gas and oil is processed and refined. Canada is the biggest exporter and China is the biggest importer of sulfur [3].

Mostly owing to the current strict global environmental regulations regarding the petroleum and gas refining processes limiting the maximum quantity of sulfur present in combustibles, the availability of sulfur has considerably grown in many countries during the last decades, so it is evident that there will be a continuous abundant supply of sulfur in the future [4]. The report from the World Business Council for Sustainable Development report, Cement Industry Energy and CO_2 Performance "Getting the Numbers Right [5], clearly indicates that global production of sulfur will continue to increase, thus assuring its continued availability.

Waste sulfur from oil processing has been stored leading to more or less controlled disposal, while waste sulfur from ore processing, in the form of gaseous sulfur oxides, is usually converted into sulfuric acid which is onwards used in the production of artificial fertilizers. Modern industrial production still requires further development of technologies that would provide incorporating waste sulfur into useful products, thus removing it from the environment. Construction materials are optimal candidates that should always be considered as potential waste acceptors in all cases of producing huge amounts of waste. Wide range of waste can be inertized by incorporating into usable and sustainable construction materials. Construction materials are preferred recipients because of large amounts of secondary sulfur and, on the other hand, because of their enormous practical application. Consuming sulfur for the production of sulfur-polymer matrix composites- concretes and mortars is a relatively new technology that has to be proven in practice. These materials are already in use, but improving their quality for further application is still needed. They have a relatively simple composition and manufacture, and interesting properties; also they continue to receive more attention since they are environmentally friendly and cost-effective. Due to their mechanical strength, fast hardening and corrosion resistance, they belong to high performance materials suitable for numerous applications, especially those in which other materials fail.

During the 1960s, there was a remarkable investment in environmental protection against sulfur deposal into the atmosphere, thus making sulfur a surplus commodity on the market, particularly in the United States of America and Canada. Therefore, extensive research programs concerning further development of using sulfur as a structural binder as well as characteristics of the obtained material including durability were initiated [1, 4].

Sulfur can completely substitute Portland cement to form new, stable, hard composite product. Usage of sulfur as a binder is based on its physico-chemical characteristics: chemical passivity, excellent resistance to aggressive agents (mainly acids and salt solutions, but not bases) and hydrophobic [6]. Unlike composite with Portland cement, composite made from sulfur requires to be heated during production. But in spite of heating, usually by the combustion of fossil fuels, composite materials made with sulfur release far fewer greenhouse gas emissions compared with Portland cement- based composite. By using sulfur as a binder, the emissions released in the calcination process during clinker production as well as the combustion emissions generated to supply heat to the calcination process are eliminated. The temperature of clinker production process is about 1,450°C [7].

Sulfur- polymer matrix composites are high performance thermoplastic materials made of mineral aggregates, filler, and modified sulfur binder, instead of cement and water as in conventional Portland cement composites at temperatures above the hardening point of sulfur (120°C). Using sulfur to make modified sulfur binder is based on its physico-chemical characteristics. According to our own terminology, the term *modified sulfur binder* means a mixture of elemental sulfur and modified sulfur-sulfur polymer. Recent experience all over the world shows that composite materials produced with modified sulfur binder instead of cement and water have significant chemical and physico-mechanical advantages comparing with Portland cement composites.

Although sulfur-polymer matrix composite is a commercial product in the United States of America, Canada, and Poland, neither its production is precisely defined process, nor its composition determined accurately. In general, beside sulfur, sulfur-polymer matrix composites contain mineral aggregates (rock, limestone, gravel, slag, dolomite, crushed concrete with Portland cement, calcium silicate, fibers of glass wool, sand, synthetic aggregates, volcanic material, ceramic material, etc.), filler (sand, fly ash, coal, dolomite, gypsum, limestone, pyrite, ground granite, calcite, barite, etc.), and various additives for polymerization or modification of sulfur, suppression of crystallization, suppression of forming and releasing sulfur gases, removing the odor of the composite material, increasing mechanical strength, etc. For a given aggregate, which is selected to be the closest to the application place, it is necessary to define the size and shape of particles, the optimal amount of sulfur in the mixture and the amount and type of additives and fillers. Production of sulfur-polymer matrix composite does not include water, which is somewhere in deficiency (the Middle East) or available but contaminated and with an excess of mineral salts, which negatively affects properties of Portland cement composite [8].

Sulfur implementation for composites production has started with using elemental sulfur as a binder. However, in spite of the excellent mechanical properties after preparation, samples showed low stability, so spalling and failure occurred after a short period of time. Namely, exposure of the material to repeated cycles of freezing and defrosting in terms of high humidity or immersion into water caused its degradation and failure. This can be explained by sulfur transformation that is occurring during the preparation of composite with elemental sulfur. On cooling of casted liquid mixture, sulfur crystallizes as monoclinic S_β at 114°C with volume contraction of 7%. Below 95.5°C, it transforms completely into rhombic S_α within 20 hours. Since S_α is more dense than S_β high stresses and micro-cracking within the material are induced. This exposes the material to moisture penetration and it fails prematurely [4, 8]. The development of modified sulfur binder contributed to better endurance of sulfur-polymer matrix composite, which focused its application for roads construction and repairing and as a building material [9-12]. Sulfur itself tends to polymerize to a large extent while chemical modification increases this tendency or prolongs the time required for the polymerization. Except the prevention of sulfur transformation from monoclinic to orthorhombic form, the degree of sulfur polymerization is increased and long chains are created due to modification. Modified sulfur has much lower thermal expansion coefficient compared with elemental one, therefore shrinkage and residual stresses upon cooling are lower [4, 13]. The polymer prevents the growth of macro- sulfur crystals. Long-term durability of modified sulfur-polymer matrix composite lies in the stability of microcrystalline sulfur [6, 14].

Contemporary tendencies desire production of materials with required characteristics based on combining constituents with various characteristics and in various proportions, as well as on the application of divers manufacturing procedures and additional material processing.

The mixture of sulfur-polymer matrix composite has to be designed to provide the best balance between mechanical strength, density and workability. A limited absorption into the solidified material will lead to good durability even in extreme conditions.

In order to obtain sulfur-polymer matrix composite with appropriate physico-chemical and mechanical properties depending on the application, the initial components should satisfy certain requirements:

- Precisely defined grain size distribution to provide maximal dense packing of aggregates grains and minimize the amount of binder;
- Spherical aggregates grains to achieve maximal dense particle packing with reduced number of voids and pores, and therefore greater strength of the material;
- High resistance to aggressive chemical agents and sudden temperature changes enabling good stability of the material;
- Low porosity in order to provide low absorption of moisture.

The function of fillers in concrete is based mainly on their size and shape. Small fraction (≤ 75 μm) of filler improves particle packing by fulfilling free space among aggregates grains, thus increasing homogeneity and fluency of mixture which leads to obtaining dense product with a minimum content of voids and pores. They provide better properties of composite, even reducing the amount of binder without strength loss. During cooling and hardening filler prevents forming large cyclic sulfur crystals thus improving mechanical properties of the composite [15-18].

Sulfur-polymer matrix composite is produced by mixing aggregate, filler and binder in the temperature range 130°-150°C. Mechanical mixers that can keep required temperature are adequate for this purpose. The proportion of aggregate, filler and binder for the preparation of sulfur-polymer matrix composites may vary depending on the application. Different procedures for sulfur-polymer matrix composite production are schematically presented in Figures 1-6 [8].

1.1. The First Procedure

Aggregate and filler are preheated and the molten modified sulfur is added, as shown in Figure 1.

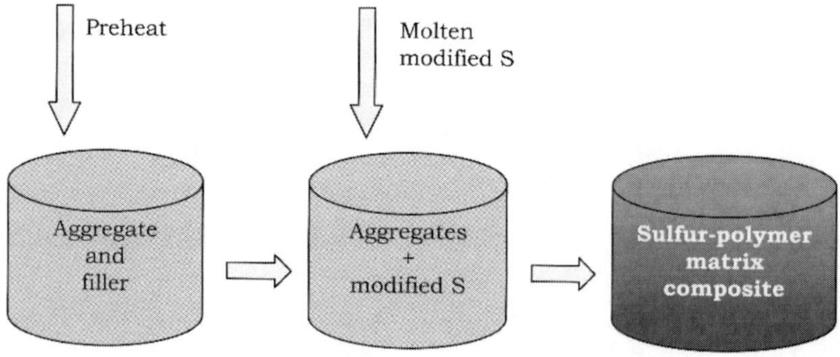

Figure 1. Early Chempruf mixing scheme.

1.2. The Second Procedure

All components are mixed at ambient temperature and then stirred along with heating until melting, as shown in Figure 2.

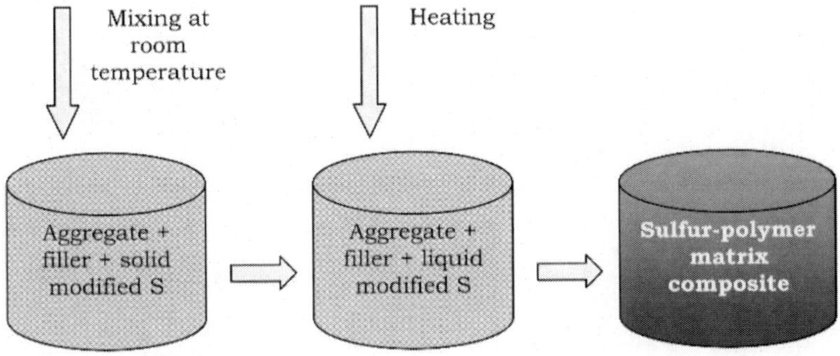

Figure 2. Sulkret mixing scheme.

1.3. The Third Procedure

Aggregate and filler are preheated and solid modified sulfur is added, as shown in Figure 3.

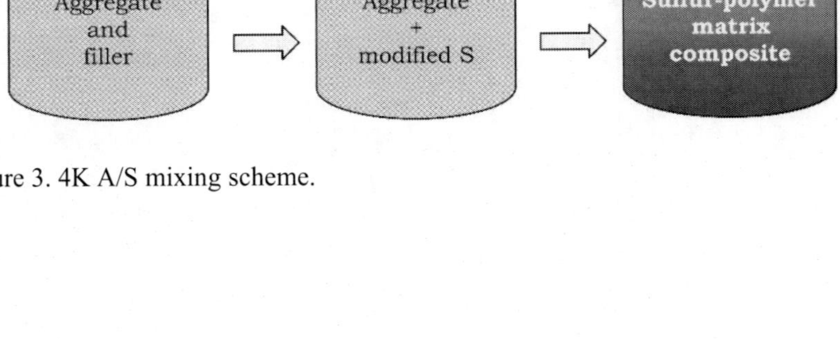

Figure 3. 4K A/S mixing scheme.

1.4. The Fourth Procedure

Aggregate is preheated and mixed with simultaneous adding filler and modified sulfur, as shown in Figure 4.

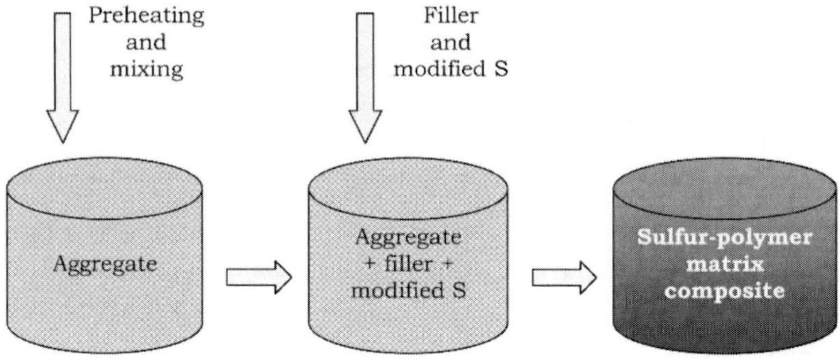

Figure 4. Later Chempruf mixing scheme.

1.5. The Fifth Procedure

Aggregate is preheated. Solid modified sulfur is added and stirred until molten state is achieved and complete aggregate is covered with sulfur. Cold filler is added in the flowing mixture, heated and stirred continuously, until the composite is homogenous, as shown in Figure 5.

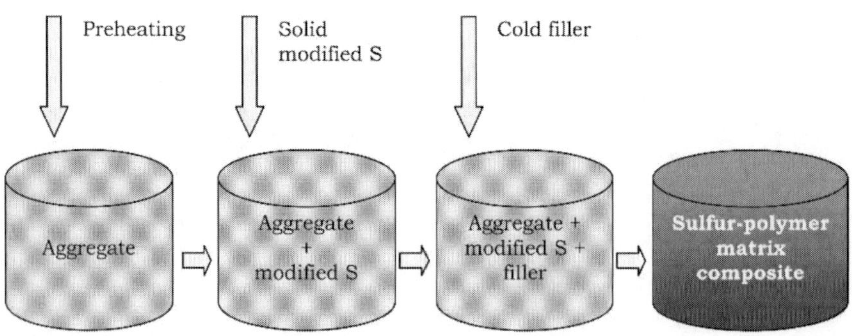

Figure 5. ACI mixing scheme.

1.6. The Sixth Procedure

Elemental sulfur and modifiers are heated at temperature 130°-150°C. Molten sulfur is mixed with modifier forming modified sulfur. Filler is preheated to higher temperature, compared with the temperature of sulfur mixing, and transferred to form sulfur cement after mixing it with modified sulfur. Aggregate is preheated similarly to filler and mixed with sulfur cement forming homogenous product, as shown in Figure 6.

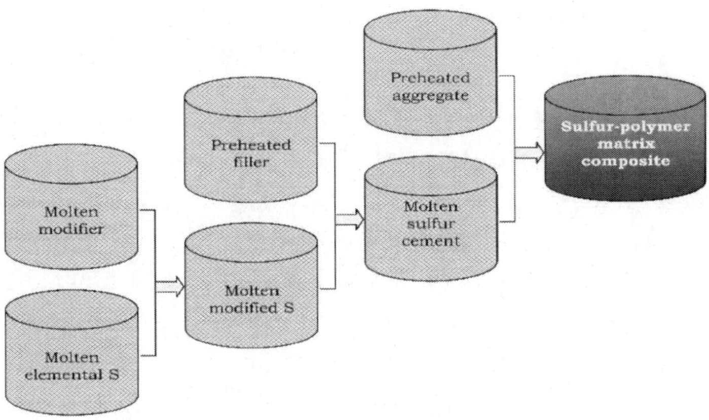

Figure 6. Mohamed and El Gamal mixing scheme.

Sulfur-polymer matrix composite can be used as a highly durable replacement for constructional materials at places within the industrial plants where the acids and salts cause premature destruction and cracking, as well as at locations with frequent cyclic freezing and melting [15, 19]. It has the most frequent usage: in food processing plants and chemical industry where aggressive materials are produced and stored, in plants for sewage treatment, on farms for domestic animals, for elements designed for sea area, drainage channels and sewage pipes, paving the pedestrian paths, curbs, sidewalks, roof tiles, railway and tramway sleepers, etc. [6]. Sulfur-polymer matrix composite can be used for solidification and encapsulation of different waste materials, such are fly ash, cement kiln dust, phosphogypsum, mercury, thus obtaining the sustainable development of construction and industrial sectors and for corrosion protection of reinforcing steel and concrete [20-24].

Since all materials during their service life are exposed to different external impacts that cause certain kind of reaction, the idea of this research was to examine the quality of sulfur- polymer matrix composite in extreme conditions, during the accelerated destruction testing in hydrochloric acid solution.

2. Experimental

2.1. Methodology of Experimental Work

At the beginning, series of experiments was performed in the laboratory conditions with the aim to conquer basic techniques and to define the order of operations in the process of mixing sulfur, aggregate and fillers, as well as to establish proportions in the consumption of sulfur and aggregate for the synthesis of sulfur-polymer matrix composite. The fine fraction of the classical building mixture of sand and gravel, as well as a variety of other materials, such as slag, ash, and coal dust were used as an aggregate. At the same time, different types of fillers were used. Application of the modified sulfur is of exceptional importance for obtaining sulfur-polymer matrix composite because it provides solidification of sulfur in open, intertwined long chains thereby preventing brittleness of the composite material. For the selection of modifier decisive factor was the market availability and price.

In this way, although the literature concerning sulfur-polymer matrix composite production already exists, our own technologies for sulfur-polymer matrix composite synthesis and for sulfur modification were developed. Therefore, our terminology differs from typical terminology available in the literature.

Modified sulfur means polymerized sulfur, which is obtained by the addition of chemical additives- modifiers to elemental sulfur.

Modified sulfur binder means a mixture of elemental and modified sulfur. By mixing it with a mineral aggregate of a suitable composition, and filler added, at sulfur melting point temperature, during solidification sulfur-polymer matrix composite is obtained.

Our previous research dealt with monitoring destruction level of sulfur-polymer matrix composite with various fillers in various aggressive environments during a period of one year. Change of mass, compressive strength and apparent porosity served as parameters for determination of destruction level [6, 25]. These examinations represented the selection experiments. Based on the results of selection experiments, the aggressive agent, as well as filler for further investigation was determined.

With the aim to assess durability, or quality of the material, in other words its life cycle, accelerated destruction was applied. This methodology is based on the influence of an agent which may cause material damage, i.e., the change of physical, chemical and mechanical properties, was applied. Accelerated destruction agent is used in increased concentrations in order to cause changes as quickly as possible.

Next step was the choice of adequate methods for monitoring changes. Since destructive test methods are not reliable for non-homogenous systems, methods based on non-destructive testing of changes in the material were selected.

Using non-destructive methods requires a standard that has well-known characteristics or characteristics which can be used for comparison with a new material.

It was quite logical to choose Portland cement composite, produced with the same aggregate and Portland cement binder, as a referent material. The reason for this is in the fact that characterizations of sulfur-polymer matrix composite and Portland cement composite are quite the same because of their similar possible applications.

Further activity was focused on defining changes in the structure of the investigated material, sulfur-polymer matrix composite, and their comparing with the referent material Portland cement composite.

Image analysis of microphotographs taken through a microscope was applied as a conventional way of these examinations, whereby the samples were cut and changes of the obtained slices were quantified. The question of the resolution necessary for observing the surface or the interior of the samples in this case was solved by using two types of microscopes, optical and scanning electron microscope (SEM).

2.2. Materials

The components used in technological procedure of sulfur-polymer matrix composite production were the following: aggregate, sulfur and modified sulfur and filler. The same type of aggregate, cement and water were used for Portland cement composite preparation.

2.2.1. Aggregate

The sand with its maximum grain size of 2 mm was used as an aggregate. It was prepared by sieving the locally available classical building mixture of sand and gravel. Chemical analysis indicated that the aggregate mainly consists of oxides of silica (89.98%), aluminum (3.61%), calcium (0.84%), iron (0.62%), potassium (0.59%), sodium (0.57%) and magnesium (0.19%).

Grain size distribution of used aggregate, determined by sieve analysis, before and after sieving, is represented in Figure 7.

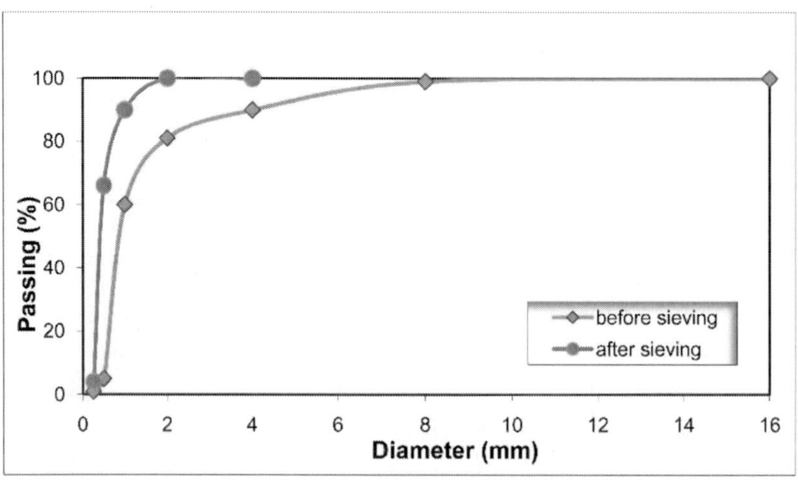

Figure 7. Grain size distribution of used sand as an aggregate.

Using aggregate with this grain size provides preparation of mortar samples, which is usual for investigation purposes.

2.2.2. Sulfur

Sulfur, which is the basic component for a modified sulfur binder, originates from oil refining process by Claus's procedure in the Oil Refinery Pancevo, and its purity is 99.9%.

According to our terminology, modified sulfur binder considers a mixture of elemental sulfur and modified sulfur- sulfur polymer.

2.2.3. Filler

The filler used in this production was alumina (Almatis, Germany). Chemical analysis indicated that the filler mainly consists of oxides of aluminum (99.3%), sodium (0.4%), silica (0.15%), potassium (0.03%), calcium (0.03%), iron (0.02%), titan (0.02%) and magnesium (0.004%).

Particle size distribution, determined by the laser diffraction using device type HELOS (H1597) & SUCELL, Sympatec, Germany, for four fillers is represented in Figure 8.

Chemical composition and densities of applied filler is given in Table 1.

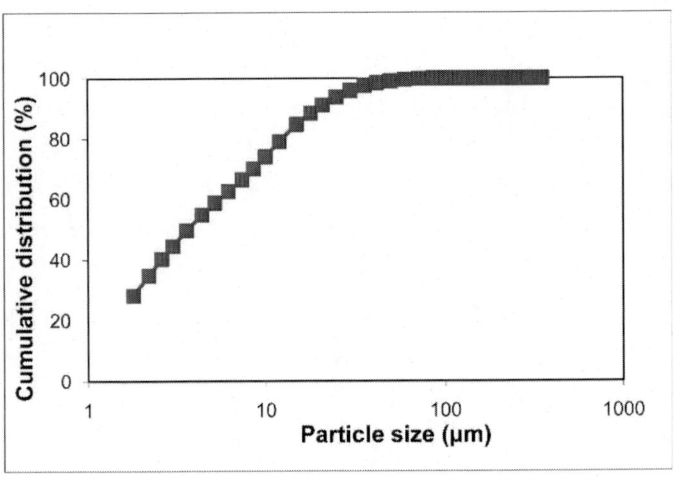

Figure 8. Particle size distribution of used alumina as a filler.

Table 1. Chemical composition and density of used alumina as a filler

Chemical composition (wt. %)										Density, ρ (g/cm^3)
SiO_2	Al_2O_3	Fe_2O_3	CaO	MgO	Na_2O	K_2O	TiO_2	P_2O_5	SO_3	
0.15	99.3	0.02	0.03	0.004	0.40	0.03	0.02	-	-	3.86

2.2.4. Cement

Chemical composition of Portland cement used in this experiment is shown in Table 2.

Table 2. Chemical composition of used Portland cement

Chemical composition (wt. %)							
CaO	SiO_2	Al_2O_3	Fe_2O_3	K_2O	Na_2O	MgO	SO_3
65.5	21.9	5.4	2.2	1.1	0.1	1.3	3.9

2.3. Polymerization of Sulfur

Sulfur modification was carried out by polymerization with cyclic hydrocarbon, dicyclopentadiene, according to the patent [26]. The procedure consisted of mixing dicyclopentadiene and melted secondary sulfur in the temperature range from 120 to 140°C and ambient pressure for 30 min, and then cooling and solidification of the obtained sulfur polymer.

Since it is very important to prepare sulfur-polymer matrix composite with modified sulfur, both elementary and the obtained polymerized sulfur were investigated by scanning electron microscope (SEM), type JEOL JSM-5800 with EDX, Figure 9, and their microstructures were analyzed according to the literature [8, 15].

The results showed that the pure sulfur, obtained by Claus's procedure consists of dense orthorhombic crystals of alpha form (S_α), Figure 9a), while, on the other hand, polymerized sulfur is composed of plate monoclinic crystals of beta form (S_β), partially polymerized in zigzag chains, Figure 9b). Therefore, it is proved that modification of sulfur is achieved.

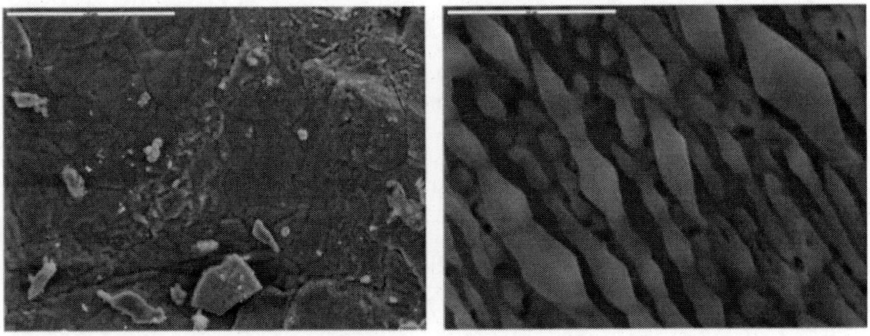

Figure 9. SEM microphotographs of sulfur: a) elemental; b) polymerized.

2.4. Manufacturing Process of Sulfur-Polymer Matrix Composite Samples

Starting points for the synthesis of sulfur-polymer matrix composite were found in procedures recommended in the literature [4, 27-30] and based on the numerous experiments, own technological procedure for sulfur-polymer matrix composite production has been developed.

In this research, sulfur-polymer matrix composite samples were fabricated according to the manufacturing technological procedure described in [6, 31]. The process included mixing both melted sulfur and modified sulfur into heated and homogenized dry mixture of aggregate and filler at sulfur melting temperature, 132–141°C. A special mixer with both temperature and stirring speed control was designed for the purpose of sulfur-polymer matrix composite production in laboratory conditions (Figure 10a)).

After homogenization and mixing that lasted for 2 min, sulfur-polymer matrix composite mixture was casted into molds preheated at 120°C and vibrated for 10 s on a vibrating table. The surface of each sample was finished and left to harden inside the molds, at room temperature. A very important characteristic of sulfur-polymer matrix composite is rapid hardening (from 15 min to several hours, depending on size and shape of the sample) that allows removal from the mold and curing in a relatively short period of time (only 24 h at room temperature). In this case, the samples were removed from the mold after 3 h and cured at room temperature for 24 h.

For the purpose of this investigation, cube- shaped samples (4 cm) were prepared.

Mix proportion for sulfur-polymer matrix composite is given in figure Figure 10b).Technological procedure for sulfur- polymer composite production is schematically presented in Figure 11.

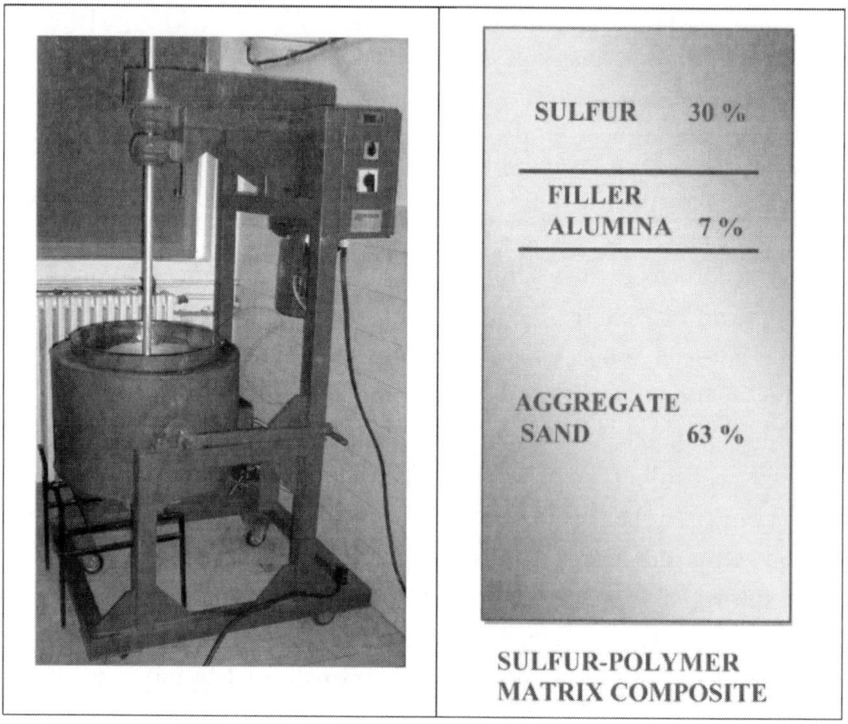

Figure 10. a) Laboratory mixer for sulfur-polymer composite production; b) Mix proportion for sulfur-polymer matrix composite.

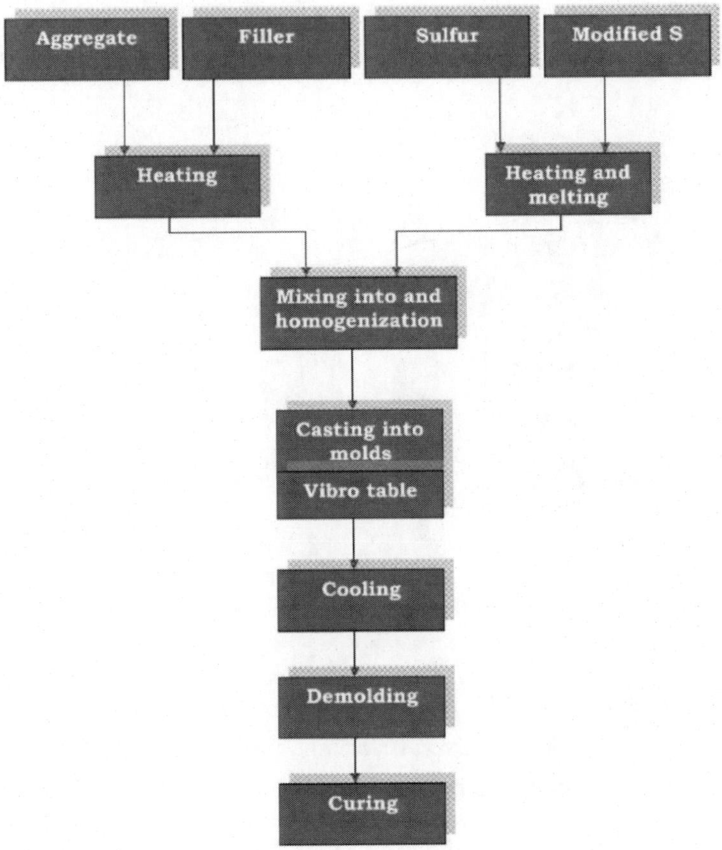

Figure 11. Scheme of technological procedure for sulfur- polymer matrix composite production.

2.5. Manufacturing Process of Portland Cement Composite Samples

In order to compare, Portland cement composite samples were made with the same aggregate as sulfur-polymer matrix composite, Portland cement and water to cement ratio w/c of 0.5. The process included mixing the aggregate with some tap water in Hobart mixer (Figure 12a)).

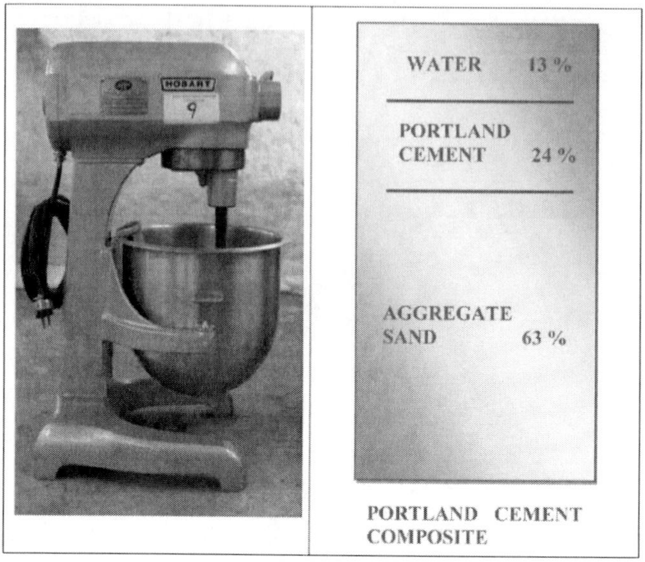

Figure 12. a) Hobart mixer for Portland cement composite production; b) Mix proportion for Portland cement composite.

Then, the cement and the rest of the mixing water were added and the mixer was reoperated. The overall mixing time was about 6 min. The Portland cement composite mixture was poured into molds and compacted by a vibrating table. The samples were removed from molds 24 h after casting and then cured in a moist room at a temperature of 20 ± 2°C with 95–98% relative humidity for additional 27 days.

For the purpose of this investigation, cube- shaped samples (4 cm) were prepared.

Mix proportion for Portland cement composite is given in Figure 12 b).

2.6. Methods

2.6.1. Accelerated Destruction

In order to evaluate quality of sulfur- polymer matrix composite, cube samples were tested under immersion in 10% HCl solution for 7, 14, 21, 60 and 180 days.

At the same time, quality of the referent, Portland cement composite samples with identical dimensions in the same aggressive solution were examined in oder to make comparison.

2.6.2. Morphology Investigation

The resulting damage of sulfur-polymer matrix composite was assessed by monitoring and analyzing morfology changes.

Microstucture changes of sulfur-polymer matrix composite samples as well as referent Portland cement composite samples during the accelerated destruction agent influence were observed using optical and scanning electron microscope (SEM).

Slices, that is cross-sections of the samples, were taken by the optical stereomicroscope with CCD (Charge Coupled Device) camera, LEICA DC 150 in oder to analyze the volume changes. The samples of sulfur-polymer matrix composite were cut into four pieces, while samples of Portland cement concrete into three slices, along the z-axis. Cutting was done by diamond cutting blade that is used for the preparation of mineralogical specimens. Software Image Pro Plus, version 6.2, with the appropriate programming procedure was used and mathematical morphological analysis was performed for the analysis of the obtained images.

In order to analyze the morphological and structural changes, sulfur-polymer matrix composite samples as well as Portland cement composite samples were taken by scanning electron microscope JEOL JSM 5800, Vega Tescan TS 5130MM.

3. Results and Discussion

3.1. Macroscopic Observation

Mutual differences in appearance of sulfur- polymer matrix composite and Portland cement composite samples subjected to the influence of hydrochloric acid solution can be detected by visual inspection.

Sulfur- polymer matrix composite samples did not show any signs of surface damage or deterioration after six months of immersion in the solution of hydrochloric acid.

On the contrary, samples of Portland cement composite exhibited a strong damage after 60 days, so further accelerated destruction testing was not possible.

Photographs of surfaces of both materials, untreated and during the accelerated destruction testing, are presented in Figure 13.

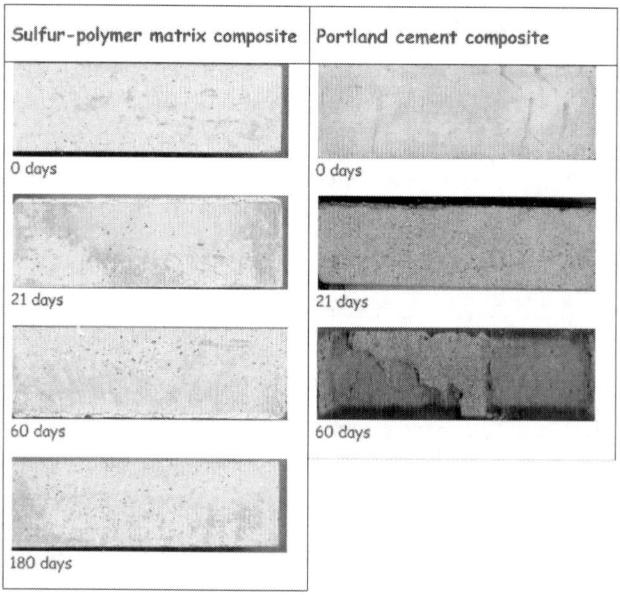

Figure 13. Surfaces of sulfur-polymer matrix composite and Portland cement composite samples during accelerated destruction testing.

3.2. Interpretation of Sulfur-Polymer Matrix Composite Resistance against Hydrochloric Acid

Sand as an aggregate and alumina as a filler, used in the production of sulfur-polymer matrix composite samples in this research, are made of mineral oxides that are mainly basic. During the hydrochloric acid solution attack on sulfur-polymer matrix composite, basic and amphoteric oxides react with the acid which results in forming metal oxides.

In sulfur-polymer matrix composite, the majority of the matrix is composed of sulfur coated materials (aggregate and filler) and sulfur accumulated in the voids between the particles. It is well known that dilute hydrochloric acid does not affect sulfur. During the acid attack, dissolving of basic and amphoteric oxides caused an increase in porosity. Acid penetration was limited to the surface and opened pores that were not coated by sulfur [6, 32, 33].

3.3. Mathematical Morphological Analysis

Images of slices obtained by a stereomicroscope were analyzed according to the selected morphological parameters which optimally presented changes in their structure.

For the processing of collected visual information software package Image Pro Plus version 6.2 was used [34].

The basis of this experiment is to determine changes in material induced by accelerated destruction. Therefore, the following morphological parameters, given in Table 3 with their symbols and descriptions, were selected.

In general, for the structure characterization it is necessary to select the (specific) parameters or group of parameters that would provide the best quantitative presentation of the structure. The point in determining values of parameters is to quantify qualitative description of the structure. A key step in quantifying the structure is the selection of morphological parameters that will give the most reliable information about the analyzed material. It is known that there are parameters that can be directly measured and parameters that can be calculated based on the previous ones.

For the characterization of the investigated structure, the most suitable parameters were: *Heterogeneity* (of the pores), share of the pores in the structure (*Hole Ratio*), *Roundness* of the pores that indicates the way of their generation and *Fractal Dimension*. Additional characterization parameters were as follows: the ratio of impact areas of the pore (*Area/Box*), as well as the ratio of the pore perimeter and its impact zone (*Perimeter* (*Ratio*)). These parameters are obtained by calculation.

In order to calculate the above parameters, it is necessary to measure the basic parameters: *Area* of the pores, *Diameter* of the pores, *Feret* and *Perimeter*.

In the structure of the investigated material two entities can be defined: a homogenous mass of the sample and the pores.

The pores may be:

- surface, formed due to one of ingredients fallout and
- bulk, formed by gas excluding during the preparation or testing of the sample.

Because of the existence of the pores in the bulk, the samples were sliced and those obtained slices were analyzed.

The pores are significant during the destruction testing because they practically represent the place where the accelerated destruction agent enters into the material, i.e., the point of the initial destruction of the material. Therefore, the characterization or quantification of pores is important for modeling material behavior as well as macro properties. On the other hand, since the sulfur-polymer matrix composite behaves as a rigid ceramic material, the pores are strain concentrators, so the shape of pores can significantly affect the way the material fractures.

The first parameter that characterizes the pores is roundness. It separates the regular pores from irregular ones. Therefore it is possible to rank them. The regular shape pores are part of the stationary fallout process whereby their creating is not affected by the structure of the surrounding material. Due to interaction with the surrounding structure, pores, which are part of non-stationary process (gas diffusion), obtain irregular shape.

Table 3. Morphological parameters [35]

Term	Graphical representation	Description
Area	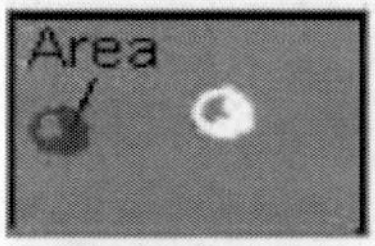	Represents the surface of the object without pores.
Area/Box	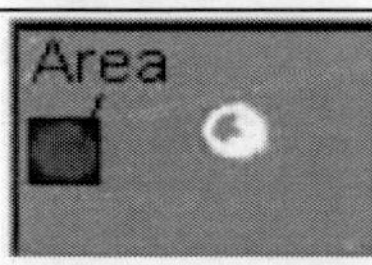	Represents the ratio of the surface of the object and its imaginary boundary surface.
Diameter (mean)		Represents the mean value of diameters measured in intervals of 2°, wherein the diameter passes through the center point of the structure with shape that differs from the circle.
Feret (mean)	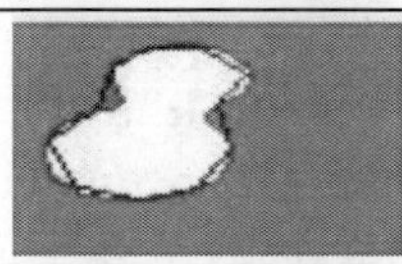	Represents the size of the chord of a circle measured at angles of 120° in relation to the initial, arbitrarily selected section.
Fractal Dimension	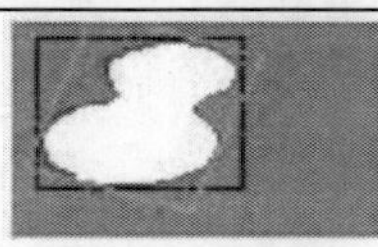	Represents the fractal dimension of the object contour.
Heterogeneity	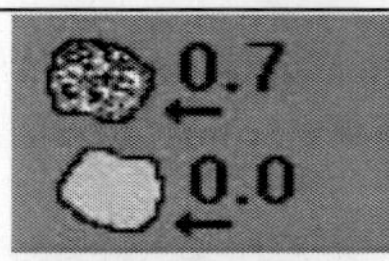	Represents share of the points that varies over 10% from the mean intensity of the object.
Hole Ratio	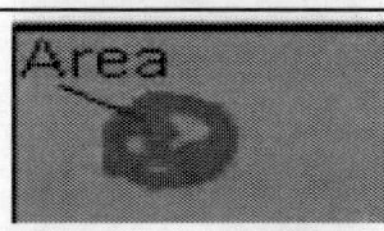	Represents the ratio of area of the object without pores and its total area, i.e., degradation level.
Perimeter (Ratio)	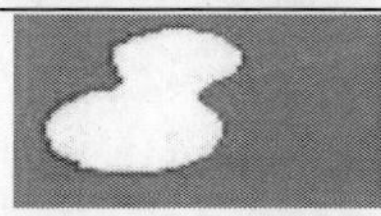	Represents the ratio of the convex contour perimeter and the contour perimeter of the object.
Roundness		Represents the curvature of an object defined by the formula: $(\text{perimeter})^2/(4 \cdot \pi \cdot \text{area})$. Round objects have roundness value 1, and the others > 1. The term *Shape Factor* is also known from the literature.

Homogeneity indicates the distribution of pores in the material. The homogeneous material has a uniform distribution of pores, meaning that pores are part of the material forming process, rather than randomly generated structure entities. The inhomogeneous distribution of pores indicates randomly generated structure entities.

Share of pores additionally describes the homogeneity and indicates the direction of change in the structure, which was significant in this study because of the destruction monitoring during the time.

One of the key parameters used for the quantification of irregularities or structure homogeneity is fractal dimension. Method for determining the fractal dimension, presenting a gradient of a measured quantity change depending on the resolution, indicates the reason for using these derived morphological characteristics. Since the slices are observed by different magnifications, determination of fractal dimension, that is rate of pore diameter change, was possible. Therefore, surface inhomogeneity for a given duration of the accelerated destruction was defined.

All parameters were analyzed in terms of their distribution in the examined structure.

Set of selected parameters- *Area/Box, Diameter (mean), Feret (mean), Fractal Dimension*, *Heterogeneity, Hole Ratio, Perimeter (Ratio)*, and *Roundness*, so all except *Area*, served for the quantification of changes in the analyzed samples. Cumulative analysis of the measured parameters changes enabled the quantification of the accelerated destruction agent effect on the material quality, that is, on the structure properties. Parameter *Area*, or measured surface, was used as a control parameter, not as structure quality criteria. The fractal dimension was determined as a measure of order of ingredients forms- components of the structure of the examined samples in order to detect possible macro changes of their morphological properties.

Apart from selecting parameters, an important element in quantification of visual information is the positioning of objects on the microphotographs of the structure. In order to ensure the adequate analysis of the images obtained by stereomicroscope, following processes were carried out: previous image processing using image equalization by best fit algorithm, removing the noise using Gauss's algorithm and at the end using modern 3D filters for clear object separation [35]. Without considering the theoretical assumptions of image- the two-dimensional distribution of radiation intensity [36], it is clear that the objects of interest are separated based on the same intensity. It means that the basis for objects determination is measuring intensity distribution which enables separating objects of interest. Sorting objects into classes based on the detected intensity is called segmentation. In the morphological analysis, the intensity distribution is called the gray scale, and process of objects separation is selection of part of a gray scale that is characteristic for the tested object. This gives a binary image, i.e., image of the objects and background image. The process of objects extraction is called binarization and it represents a separation of structural elements according to the detected classes. In this way, objects for analysis are extracted, that is the pores are separated from the resting part.

At the end, selected morphological parameters were measured.

Scheme in Figure 14 presents the quantification of the structure properties.

In order to automate and accelerate the analysis, all the procedures have been put into a special program known as "macro" in programming practice.

In this research, four slices on both sides with two positions per side of each sulfur-polymer matrix composite sample were analyzed, representing 16 images per sample for processing. In the case of Portland cement composite samples, three slices of each sample were made; on both sides two positions per side were analyzed, which represented 12 images per sample for processing.

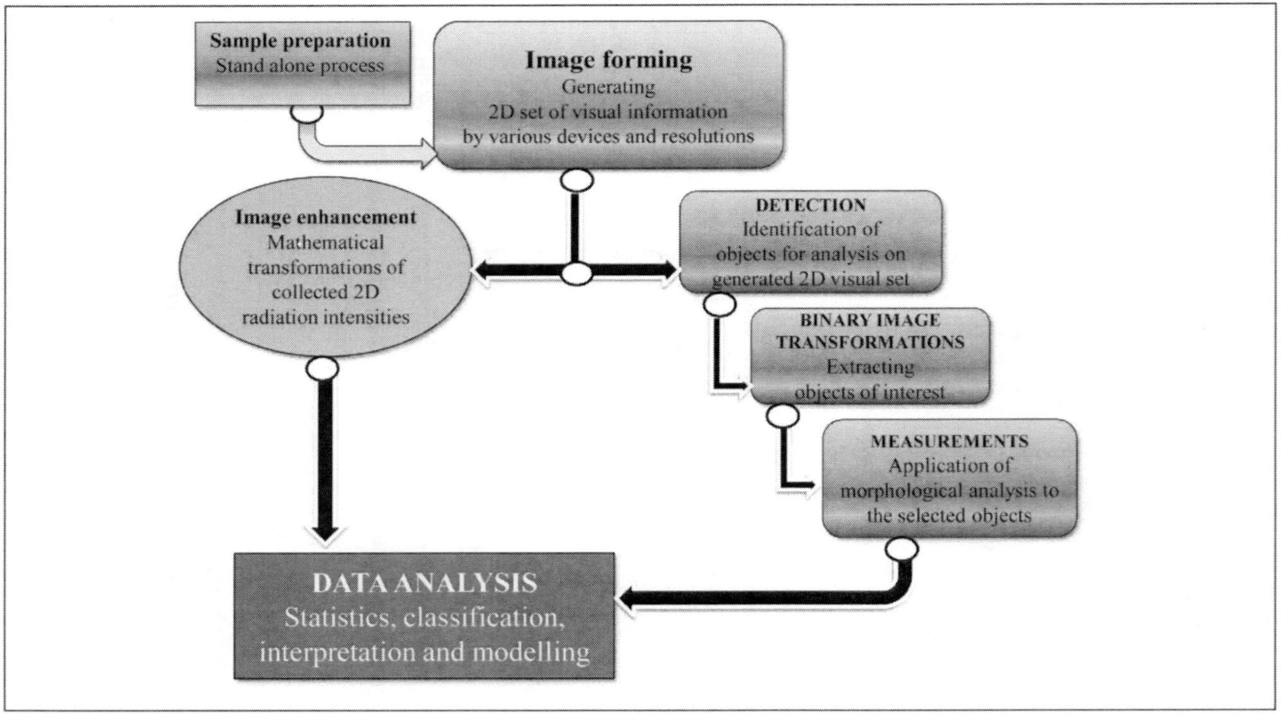

Figure 14. Quantification of the structure properties [25].

Examples of the obtained distributions of measured values for the selected parameters are given in Figure 15.

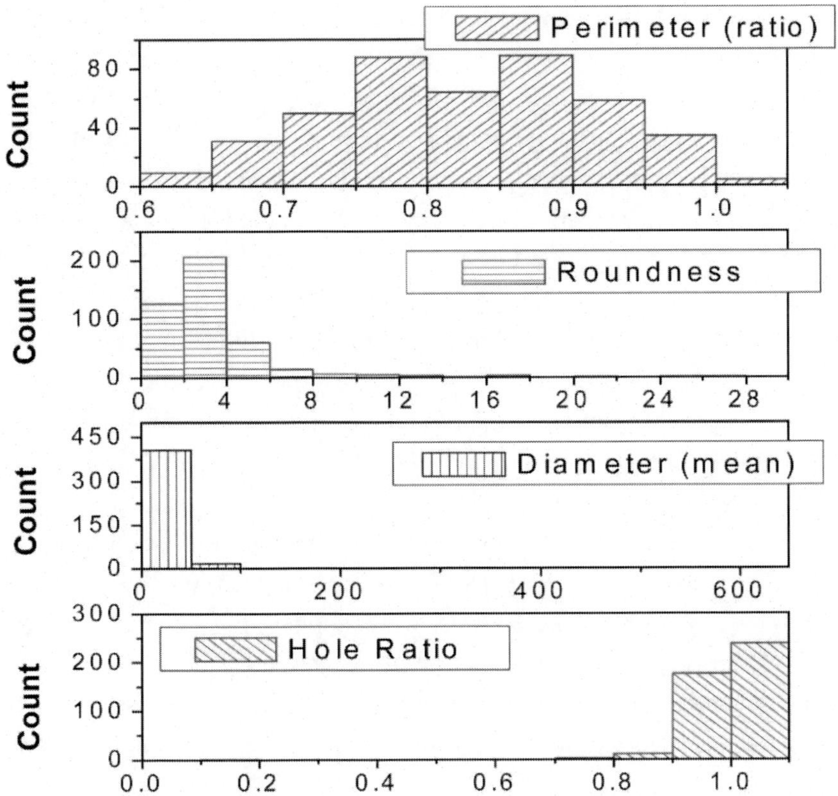

Figure 15. Examples of the obtained distributions of the measured values for the selected parameters.

The problem that arises when analyzing the results of the structure quantification is the selection results which will be presented along with the presentation method. Having in mind the aim of this investigation, it is clear that the results given in their original form at Figure 15 are illustrative, and during the analysis it was necessary to use higher level of processing these results, a complex statistical analysis. Since the essence of the study was monitoring structure changes in time, the application of statistical parameters for comparing changes and identifying their character was a logical step. The quality of statistical analysis was achieved by great number of individual measurements of the tested material structures. Therefore, in the following text, the focus is placed on the interpretation of the change described by statistical values of parameters.

Morphological measurements generally provide a large number of values that need to be analyzed. The first step is called data reduction based on the measuring range, where all the results beyond defined limits for the selected morphological parameters are discarded. For the quality analysis and classification of the measured data (called data reduction procedure) [35], measured values of the parameter *Area*- surface of the structure object were used. Usage of this parameter was optimal because it provides the easiest way to determine whether it is the structure object or the structure damage. Lower limit of the measured surface was the limit of resolution of the stereomicroscope used to generate images. On the basis of this morphological parameter first reduction of measured objects for further analysis was performed.

It is clear that the statistical methods were the key for data processing in the first step of the analysis. Classical statistical method of data processing and correlation analysis were applied here because it was essentially most important to determine the variations depending on the time and penetration depth of the agent of accelerated destruction. The aim of this analysis is correlating these results with previously obtained models of material behavior.

Two types of statistical data processing were performed here.

The first analysis is the analysis of all measured parameters in a fixed time for each sample.

The second analysis is comparing time change of measured morphological parameters for each sample.

In the first case, the influence of the accelerated destruction agent on the bulk change in the structure, quantified by selected measured morphological parameters, was analyzed. In the second case, the temporal dependency of the selected parameters change along the sample depth was analyzed.

3.3.1. Analysis of the Results Obtained for Sulfur- Polymer Matrix Composite

The surfaces of the slices obtained by cutting samples of sulfur-polymer matrix composite are taken by a stereomicroscope, and typical images are shown in Figure 16.

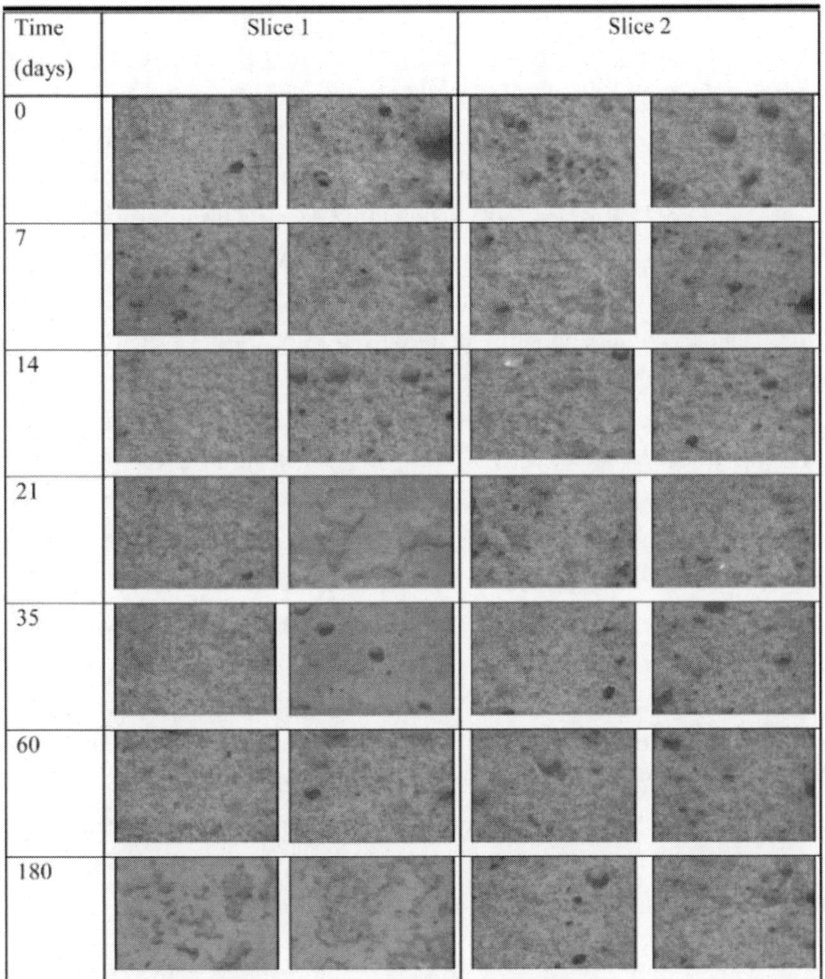

Figure 16. The stereomicroscope images of sulfur- polymer matrix composite samples during accelerated destruction testing.

For the processing of collected visual information was used software package Image Pro Plus version 6.2 and the above described program procedure were used.

After data reduction, extensive mathematical analysis using statistical procedures was performed for each sulfur- polymer matrix composite and Portland cement composite sample as well as time analysis of all measured parameters.

First, statistical analysis was performed to define the homogeneity of the sample. In this analysis, it was important that the values of measured parameters on the slices of a certain sample do not vary significantly. It means that the inhomogeneity shows whether a sample is not affected by the influence of the agent of accelerated destruction throughout the volume, or does not respond. Based on ultrasonic velocities results, i.e., analysis in the bulk, it was found that significant changes in sulfur-polymer matrix composite samples occurred after 21 days of treatment- a homogeneity change [37-39]. This is in correlation with the measurements of surface changes on the same samples [33, 40]. Having in mind these results, the question is whether a complete penetration of the accelerated destruction agent into the material occurred between 14 and 21 days and whether after that period an entire volume interaction of between the sample and the accelerated destruction agent existed.

Mean values of measured parameters during the treatment time, Figure 17, indicate that there is a homogeneity change, meaning that more significant interactions took place than in the first 21 days. In this way, the impact of sulfur on the material behavior in terms of inhibition of the observed changes in the structure can be discussed.

Figures 18-20 present the homogeneity comparison between sulfur-polymer matrix composite samples treated for 0, 21 and 180 days on the basis of mean values of the parameters measured on the slices.

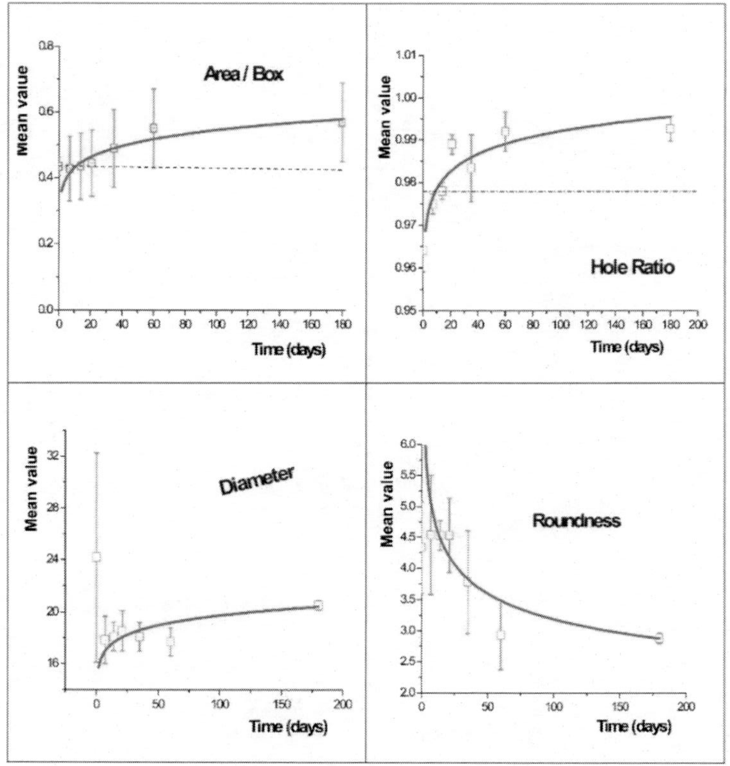

Figure 17. Mean values of measured parameters- homogeneity change of sulfur- polymer matrix composite during accelerated destruction testing.

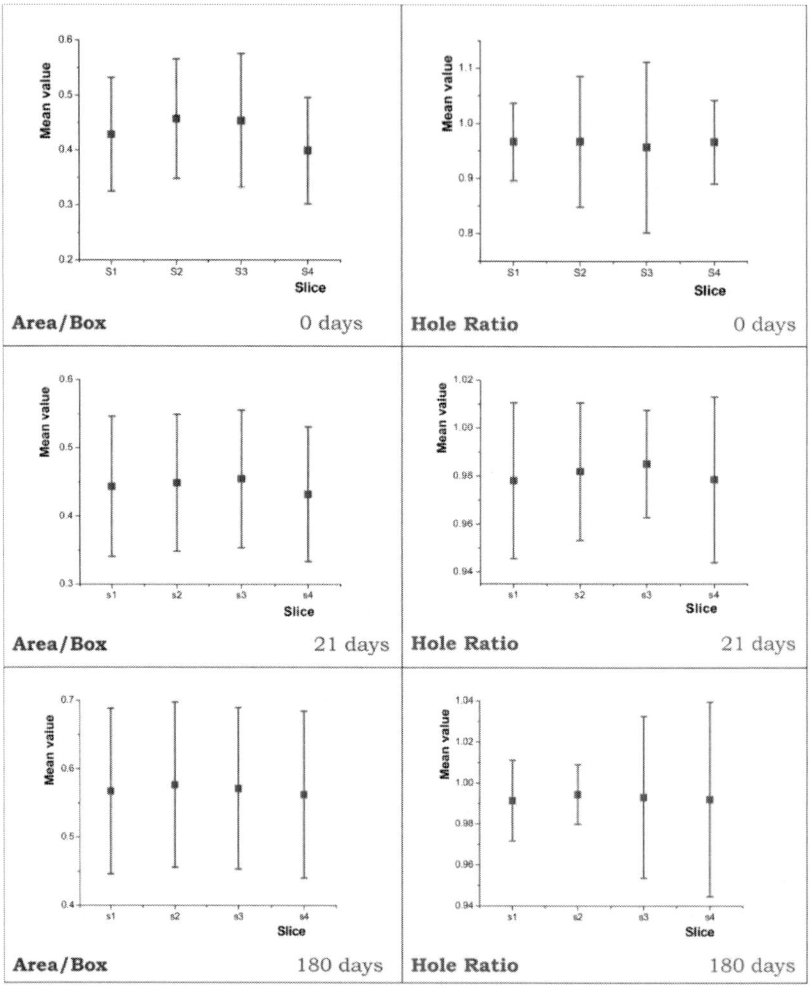

Figure 18. Mean values of parameters *Area/Box* and *Hole/Ratio* on slices of sulfur- polymer matrix composite samples after 0, 21 and 180 days of accelerated destruction testing.

Homogeneity parameter is complex, and therefore the idea was to analyze the change of structure homogeneity through the volume from various aspects, that is by several parameters. Unlike surface analysis, where the homogeneity is measured by the selected parameter distribution, homogeneity change throughout the volume must be observed by a number of parameters.

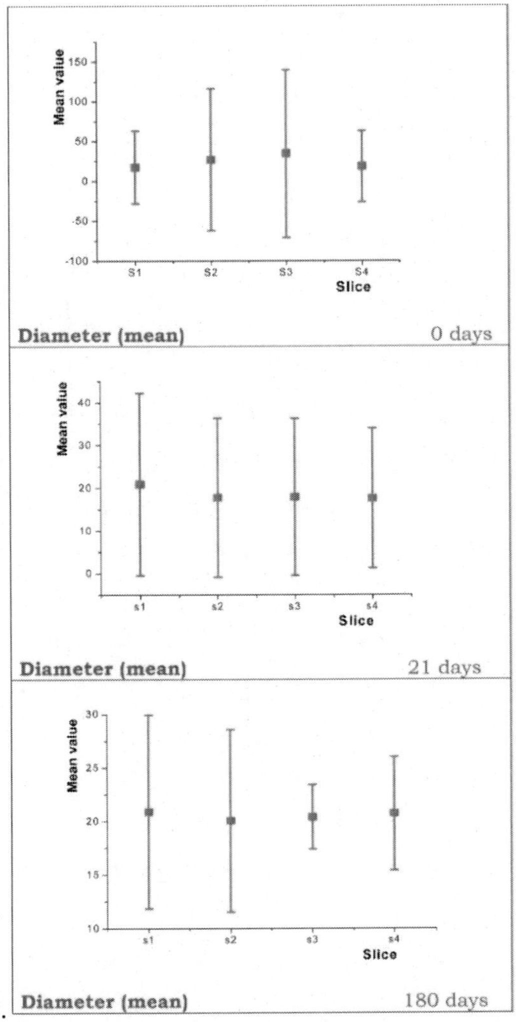

Figure 19. Mean value of parameter *Diameter (mean)* on slices of sulfur- polymer matrix composite samples after 0, 21 and 180 days of accelerated destruction testing.

The trends of selected parameters changes were analyzed through the samples volume. Variations of mean values of the measured parameter, wherein the standard deviations have similar values, can be observed on the presented graphs. Untreated samples exhibit the greatest variations of the parameters mean values, and they reduce by increasing the treatment time. On the other hand, the standard deviations, or ranges of measured parameters values remain almost the same until the end of the treatment time. Therefore, it can be concluded that a homogenization of the sample occurred during the treatment time.

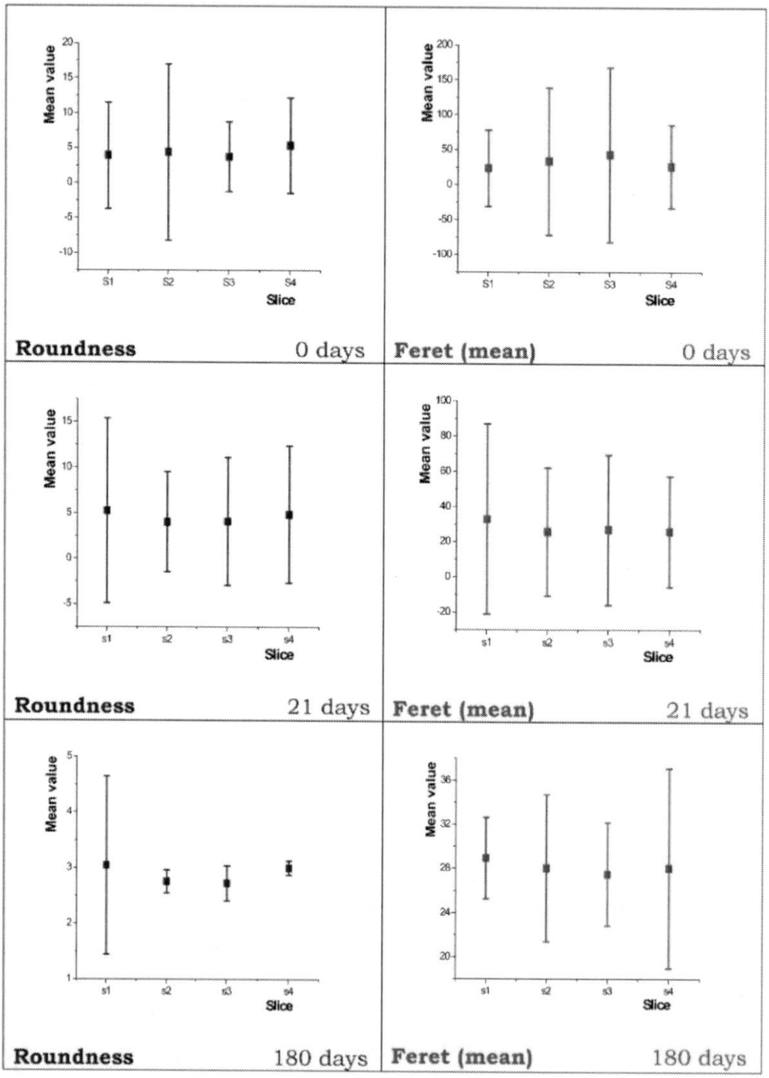

Figure 20. Mean values of parameters *Roundness* and *Feret (mean)* on slices of sulfur-polymer matrix composite samples after 0, 21 and 180 days of accelerated destruction testing.

Based on the given analysis, it may be concluded that the greatest homogeneity increase takes place at 21st day of treatment which is in accordance with the results of ultrasonic measurements [37-39]. The morphological analysis is interesting after 21 days of treatment because of evident greater volume changes of certain parameters which positively or negatively affect the homogeneity achieved at approximately 21st day of accelerated destruction agent influence. However, regardless the effect, the action of the accelerated destruction agent was towards rearranging sulfur, which was the main moderator of homogeneity changes.

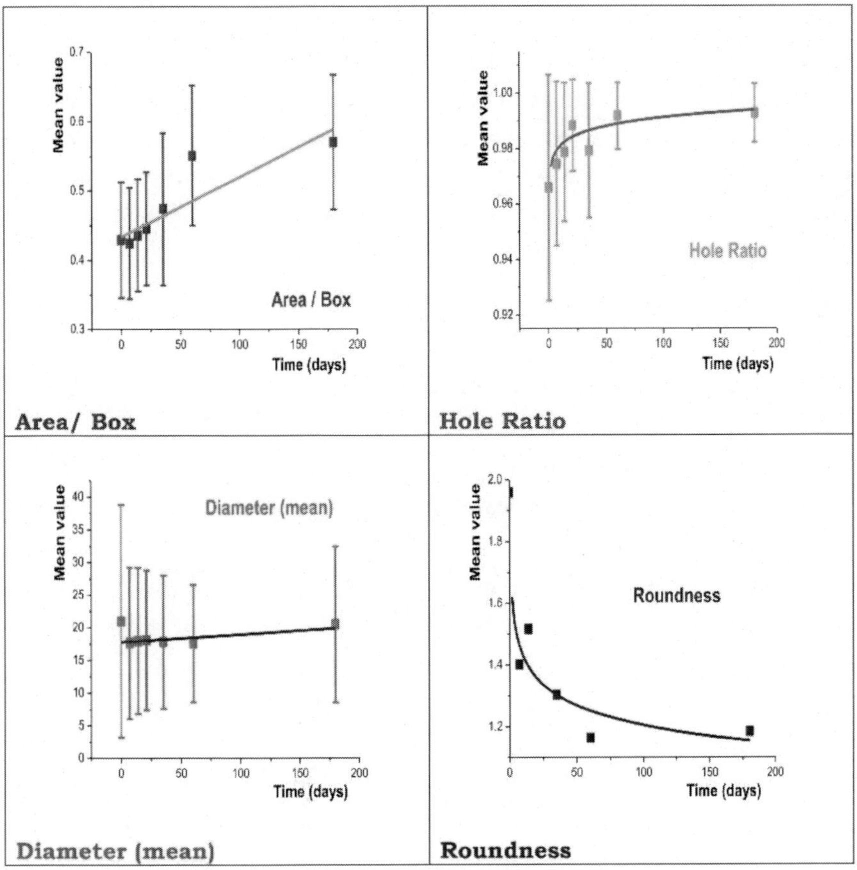

Figure 21. Change of measured parameters through the volume of sulfur- polymer matrix composite samples during accelerated destruction testing from 0 to 180 days.

Further analysis was the analysis of time changes in cumulative parameters for each sample. For this analysis, the most important is whether the type of changes is the same or different compared with the functions obtained by surface examination and ultrasonic measurements (the power functions) [33, 37-39].

Figure 21 shows the graphs for four measured parameters through the volume of sulfur- polymer matrix composite sample during the treatment time from 0 to 180 days.

As it can be seen from the graph shown in Figure 21, the time change kept the same form- power function. Compared with the surface analysis, a lower correlation degree is present here. This is entirely in accordance with the supposed model, since, unlike surface changes, bulk changes are cumulative changes of higher interaction material- agent, which results in a lower correlation degree between the examined parameters. However, it is important that the working model remains the same and that the primary interaction, referring to the interaction agent- binder, is unchanged.

3.3.2. Analysis of the Results Obtained for Portland Cement Composite

Analysis of Portland cement composite, as a reference material, was performed in the same manner.

The results of the mean values of parameter *Area/Box* measured on samples slices and for used times of treatment by the agent of accelerated destruction are given in Figure 22.

As seen on Figure 22, homogeneity of the Portland cement composite samples obviously decreases, since the standard deviation of the presented parameter is great. This phenomenon is expected because it was experimentally found that after 21 days of the treatment the material practically lost its initial properties.

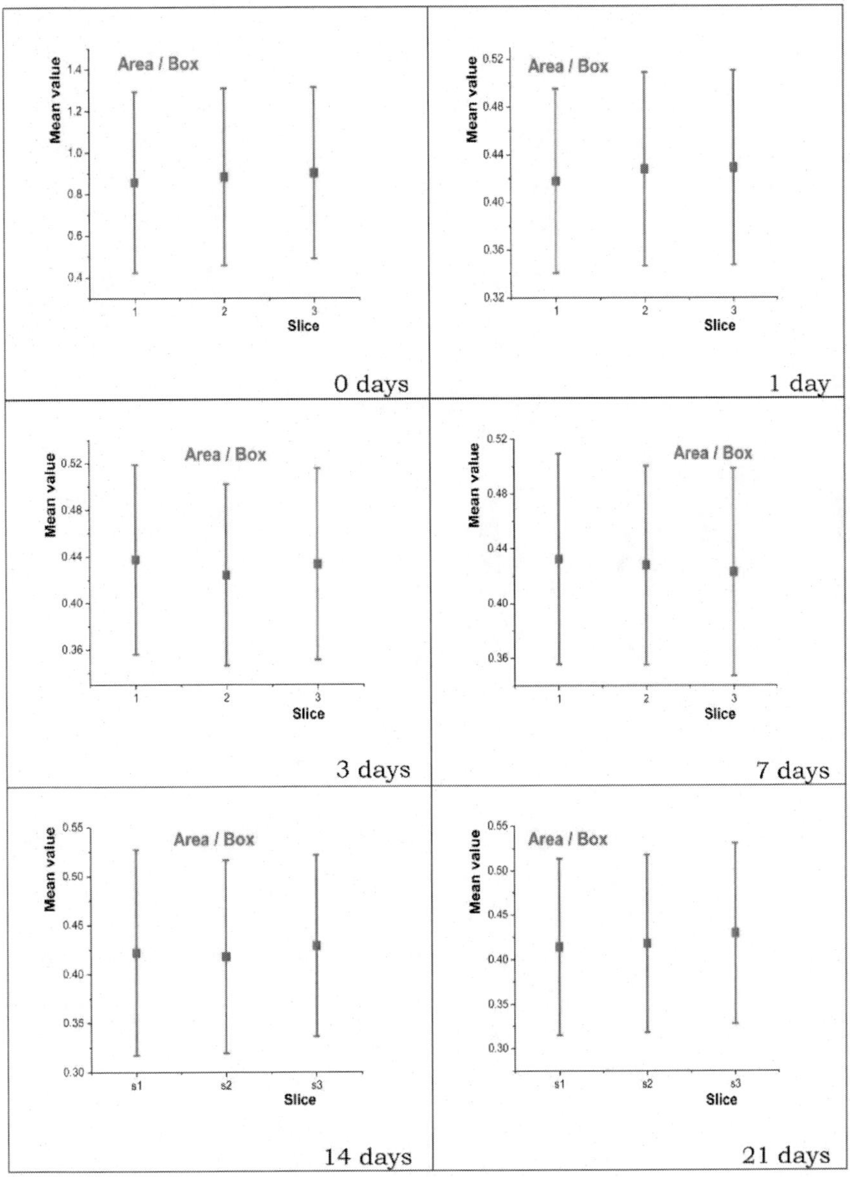

Figure 22. Mean values of parameter *Area/Box* on slices of Portland cement composite samples during accelerated destruction testing from 0 to 21 days.

Time changes of the examined parameter *Area/Box* are shown in Figure 23.

Based on this analysis, it can be seen that the influence of accelerated destruction agent leads to faster destruction of Portland cement composite structure, in other words leads to the change in homogeneity.

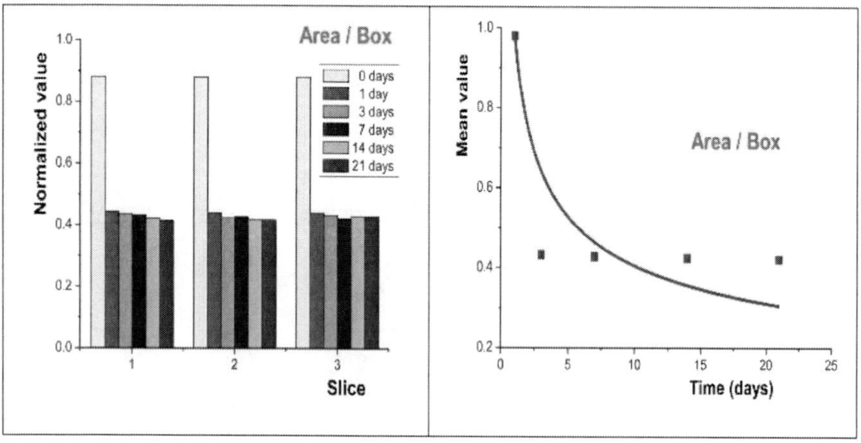

Figure 23. Time dependence of parameter *Area/Box* for Portland cement composite samples during accelerated destruction testing from 0 to 21 days.

3.3.3. Comparison of the Results Obtained for Sulfur- Polymer Matrix Composite and Portland Cement Composite

Since graphs of the parameters changes given within the above results (Figures 17-23) are illustrative and there was no need for their detailed explanation, the overall interpretation of changes in the structures of both materials will be presented in the following text.

Figure 24 shows a comparative view of mean values of parameter *Area/Box* on slices of sulfur- polymer matrix composite and Portland cement composite samples after 21 days of treatment, while mean values of the same parameter during the entire accelerated destruction time for both materials are compared in Figure 25.

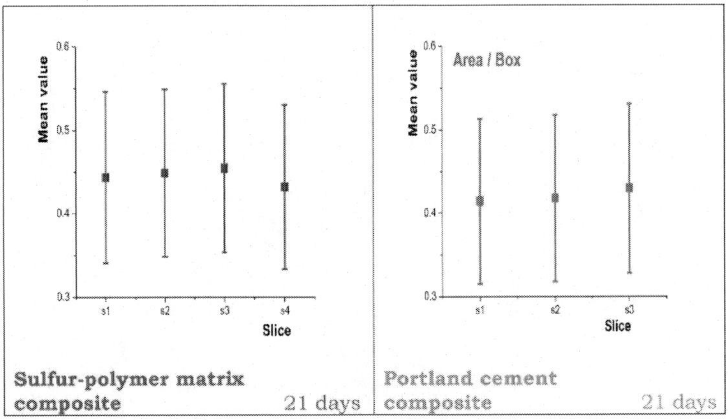

Figure 24. Comparison of mean values of the parameter *Area/Box* per slice of sulfur-polymer matrix composite and Portland cement composite sample after 21 days of accelerated destruction testing.

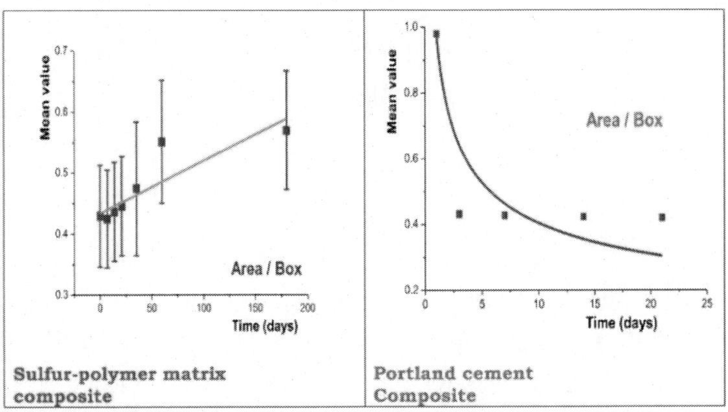

Figure 25. Comparison of mean values of the parameter *Area/Box* for samples of sulfur- polymer matrix composite and Portland cement composite during the entire time of accelerated destruction testing.

Based on these results it is evident that the examined materials react differently to the action of the accelerated destruction agent. Sulfur-polymer matrix composite exhibited noticeable changes in resistance against accelerated destruction agent impact. This can be explained by rearranging the structure and by increased influence of sulfur as a binder on the properties of sulfur- polymer matrix composite after a certain time.

In the case of Portland cement composite, the existence of the initial material is questionable after 21 days of treatment. Changes in both systems were observed after 21 days of treatment. It can be assumed that after this period the complete penetration of accelerated destruction agent through the volume of the samples took place.

In case of sulfur- polymer matrix composite it is assumed that a full penetration of accelerated destruction agent leads to its interaction with sulfur throughout the entire volume, and that this interaction intensifies binding properties of sulfur which contribute to compacting the structure and better resistance of the material against destruction.

A redistribution of sulfur through the sulfur- polymer matrix composite volume also occurs, and it practically minimizes changes during the time. That was observed in experiments where sulfur- polymer matrix composite samples with various compositions did not exhibit significant changes even after a year of aggressive treatment [6, 31, 41].

For Portland cement composite, the action of the accelerated destruction agent is destructive and totally destroys the original structure of the material.

3.4. Morphological Analysis by SEM

It is interesting that drastic qualitative and quantitative changes are occurring in both materials after 21 days of accelerated destruction agent influence. According to the fact that the materials testing is related to defining or quantifying the structure that is homogeneity of the material, the following logical part was investigation of the structure with higher resolution. For this purpose, a scanning electron microscope was used. The aim of the applied SEM analysis was to determine morphological changes during the time in terms of the sulfur distribution.

The testing methodology included monitoring the structure changes of sulfur- polymer matrix composite before the treatment and up to six months of the accelerated destruction agent action. These structures were compared with the structures of Portland cement composite before the treatment and treated up to 21 days.

Microphotographs generated by the SE detector were analyzed in order to determine the morphological structure of the samples. Microphotographs obtained by the BSE detector were used to provide information of the morphological characteristics of structure components in the samples. To determine the final distribution of sulfur in the structure, element mapping, which means observing a sample in the energy range of the selected element, was applied.

3.4.1. SEM analysis of Sulfur- Polymer Matrix Composite

SE microphotographs of sulfur- polymer matrix composite samples are presented in Figure 26.

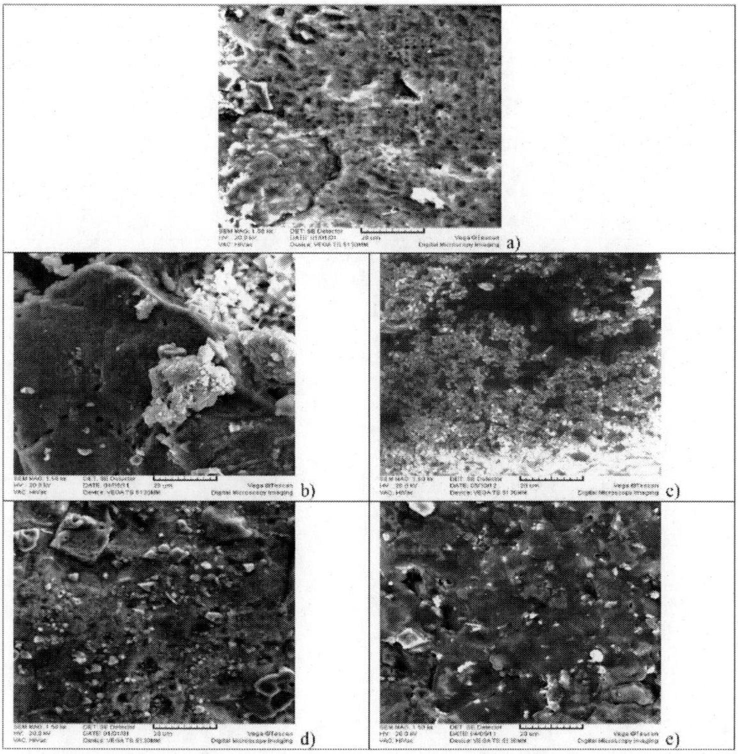

Figure 26. SE microphotographs of sulfur- polymer matrix composite samples during accelerated destruction testing: a) 0 days, b) 14 days, c) 21 days, d) 60 days, and e) 180 days.

As seen on SE microphotographs, the existing binding phase, sulfur, and separated aggregate grains can be distinguished.

Clearly defined aggregate and sulfur binder that looks like a monolithic integral part of the system can be noticed in the structure of the untreated sulfur- polymer composite sample, Figure 26 a).

Sulfur- polymer matrix composite samples exposed to the accelerated destruction agent influence show more pronounced monolithic structure. Also, ‘‘drowning’’ of the aggregate into the binder phase- sulfur, homogenization, can be observed. With increasing the treatment time, the structure exhibits further changes in the direction of homogenization and ‘‘drowning’’ of the aggregate in the sulfur binder compared with the initial structure, Figure 26 b), c), d). Based on the analysis of the structural changes, secondary homogenization of sulfur- polymer matrix composite treated by acid, with clearly visible secondary binding of the aggregate and primary binder phase, expressed after 180 days, is evident, Figure 26 e).

BSE microphotographs of sulfur- polymer matrix composite samples are presented in Figure 27.

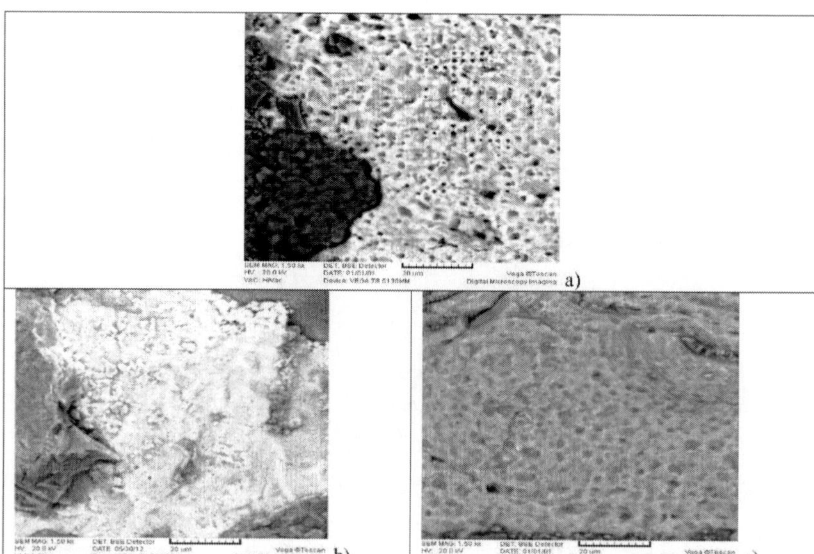

Figure 27. BSE microphotographs of sulfur- polymer matrix composite samples during accelerated destruction testing: a) 0 days, b) 21 days, and c) 180 days.

The obtained map for sulfur and the corresponding BSE microphotograph of sulfur-polymer matrix composite sample is presented in Figure 28.

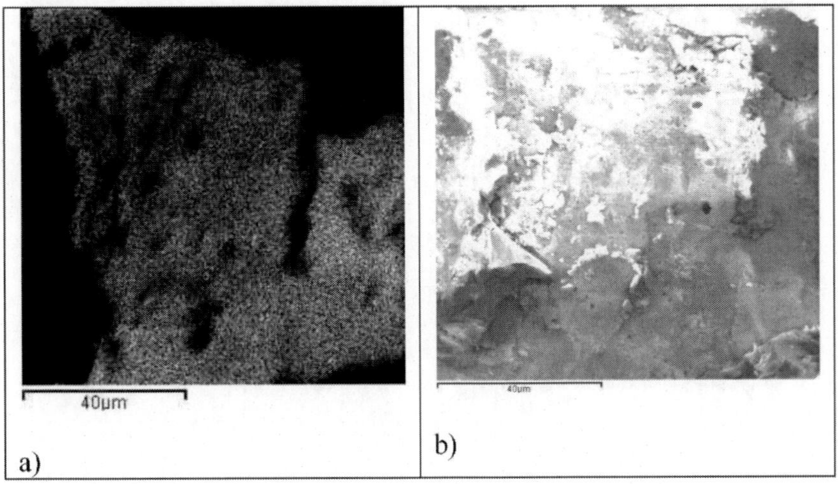

Figure 28. a) Map of sulfur and b) BSE microphotograph of sulfur-polymer matrix composite for 21 days of accelerated destruction testing.

As seen in BSE microphotographs, Figure 28, structural elements can be differentiated; the binder phase- sulfur is clearly distinguished. Segregation of sulfur in sulfur- polymer matrix samples is evident and therefore the assumption that the accelerated destruction agent provokes the secondary homogenization induced by the chemical activation of sulfur is confirmed.

According to the sulfur map in Figure 28 a) initial sulfur segregation was obviously taking place after 21 days of the accelerated destruction agent action which indicated a concentration of sulfur, as well as rearranging the ingredients within the entire volume of the sample. This meant structure changes as a result of an external impact- chemical agent. The rearrangement begins after a certain time of the accelerated desruction agent action. It occurs only when the interaction with the agent is achieved in the entire volume of the sample. It is important to mention that this segregation is only an indicator of sulfur rearrangement within a structure to increase its homogeneity.

3.4.2. SEM Analysis of Portland Cement Composite

SE microphotographs of Portland cement composite samples are given in Figure 29.

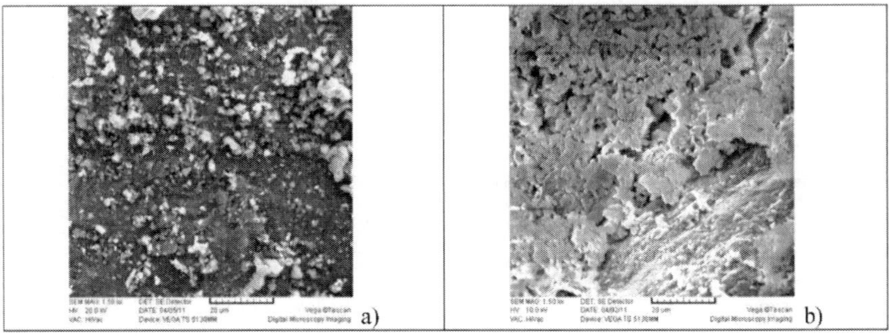

Figure 29. SE microphotographs of Portland cement composite samples during accelerated destruction testing: a) 0 days, b) 21 days.

As seen on Figure 29 a), components- aggregate and Portland cement binder are clearly distinguished in homogeneous grained structure which is similar to the structure of untreated sulfur- polymer matrix composite samples.

After 21 days of acid treatment, Figure 29 b) qualitative change can be observed which means that agglomeration and production of plate formations on the sample surface are occurring. In terms of structural changes, the chemical agent by acting on components of Portland cement composite provoked creating plate instable structures, which led to complete degradation as a macro effect.

3.4.3. Comparison of the SEM Results Obtained for Sulfur- Polymer Matrix Composite and Portland Cement Composite

SE microphotographs of sulfur- polymer matrix composite and Portland cement composite samples after the same treatment time- 21 days are given in Figure 30.

According to microphotographs in Figure 30, structural differences between sulfur- polymer matrix composite and Portland cement composite after 21 days of acid treatment are evident. Accelerated destruction agent provoked changes on micro scale in the structure of both materials.

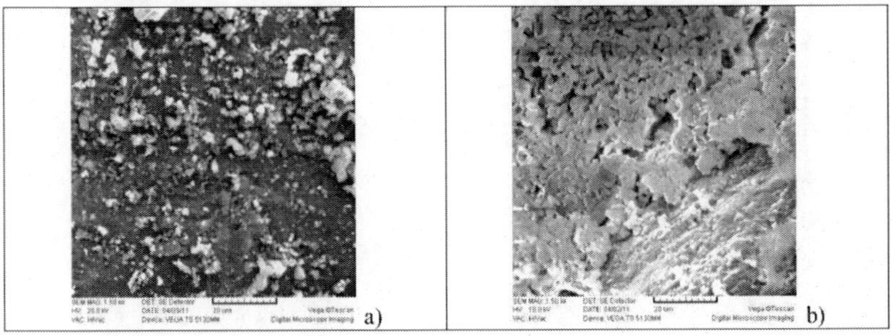

Figure 30. SE microphotographs of samples for 21 days of accelerated destruction testing: a) sulfur- polymer matrix composite, b) Portland cement composite.

The analysis of SEM microphotographs of sulfur- polymer matrix composite and Portland cement composite samples pointed out that the investigated materials exhibited quite the opposite responses to changes caused by the accelerated destruction agent action.

In the case of sulfur- polymer matrix composite, binder phase- sulfur rearranged and conditionally homogenized the structure. Secondary bonding of the aggregate, which additionally homogenized the material, can also be noticed. It can be concluded that chemical activation of sulfur-polymer matrix composite, precisely sulfur as a binder, resulted in structural changes of the material. On the contrary, unlike sulfur- polymer matrix composite, the original, untreated structure of Portland cement composite was completely degraded, whereby the binder phase was destroyed owing to the accelerated destruction agent influence [1].

CONCLUSION

In order to successfully use secondary sulfur as a constituent of certain materials, it was necessary to clearly define the technological process of the material preparation, as well as the area of its application.

Synthesis of sulfur-polymer matrix composite was performed by mixing molten sulfur and modified sulfur at melting point temperature of sulfur into dry heated mixture of aggregates and fillers. Mixing lasted until forming a homogeneous viscous mixture which was casted into preheated molds and vibrated. Modification of sulfur in this study was carried out by mixing the molten sulfur with cyclic hydrocarbon, dicyclopentadiene (DCPD).

In order to assess quality of sulfur-polymer matrix composite accelerated destruction method was applied. 10% hydrochloric acid solution was chosen as an accelerated destruction agent.

In this study, the focus was put on modern methods and methodologies for quantifying the materials properties.

The basis of quantification was morphological analysis of the material structure. The samples were cut in slices which were analyzed in order to define the way of accelerated destruction agent impact throughout the entire volume of the material.

Examinations were directed towards choosing structure parameters that would adequately represent changes throughout the volume of the material structure. Since the previously performed ultrasonic testing indicated a homogeneity change after a certain period of exposure to accelerated destruction agent influence [37-39], the focus of this investigation was placed exactly on the analysis of changes in homogeneity during time.

According to the trends of selected parameters variations throughout the samples volume, it was concluded that a homogenization of sulfur-polymer matrix composite sample occurred during the time of accelerated destruction. The greatest homogeneity increase happened at 21st day of treatment. The influence of the accelerated destruction agent was towards rearranging sulfur, which was the main factor of homogeneity changes. On the contrary, according to the similar analysis, homogeneity of the Portland cement composite samples decreased. Changes in both materials were observed after 21 days of the agent action. After this period it can be assumed that the complete penetration of accelerated destruction agent occurred through the volume of the samples. Based on this mathematical morphological analysis, it can be concluded that the materials responses to the influence of the accelerated destruction agent were mutually opposite. This can be explained by structure rearranging and by the enhanced influence of sulfur as a binder on the properties of sulfur- polymer matrix composite after a certain time. After 21 days of treating Portland cement composite samples, the existence of the initial material is questionable.

It was possible to show sulfur- polymer matrix composite structure rearranging by observing the material structure with higher resolution and by choosing another method that would indicate the volume changes. The scanning electron microscopy was applied as a highly resolution method. This method clearly identified changes- rearrangement of the structure during the time.

Namely, significant changes in resistance of sulfur- polymer matrix composite samples towards the influence of the accelerated destruction agent during the treatment time can be noticed. Explanation for this can be found in rearranging the structure and increased influence of sulfur as a binder on the properties of sulfur- polymer matrix composite after certain period. During penetration of chemical agent, its interaction with sulfur throughout the volume is occurring. Owing to this interaction, bonding properties of sulfur are increased, which lead to compacting of the structure and the material becomes more resistant to destruction. Redistribution of sulfur throughout the sample volume is also taking place, practically minimizing changes during the time.

SEM results also pointed out that the homogeneity of sulfur- polymer matrix composite changed, but after 21 days much slower because of the slower structure rearrangement. It means that after 21 days of treatment, the effects of the accelerated destruction agent on the material were substantially weakened. One of the used methods, mapping, showed that sulfur was concentrated in some locations. It is clear that sulfur concentration leads to improving its binding efficiency, which at the macro level presents increased homogeneity and improved mechanical properties. From the structural point of view, this means that the accelerated destruction agent action leads to rearrangement and that a key player that realizes rearranging was sulfur. The existence of starting Portland cement composite is questionable after 21 days of treatment. The material was degraded, whereby binder phase- Portland cement was totally destroyed.

The presented SEM analyses using specific methods of investigating materials supported the results and conclusions arising from mathematical morphological analysis.

According to the presented research, it can be concluded that the methodology of accelerated destruction can be used to provoke changes in the structure of materials and thus properties changes. The methods used for this investigation adequately demonstrate the possibility of detecting accelerated destruction influence on the sulfur- polymer matrix composite structures.

The results of previous investigations demonstrated the possibility of using sulfur- polymer matrix composite as a corrosion resistant material. The material exhibited excellent mechanical properties compared with those of Portland cement composite and showed no signs of strength loss or degradation during the exposure to accelerated destruction agent [42-44].

The presented analysis proves that for both materials changes in the structure on micro scale exist and they are result of interactions with the accelerated destruction agent, which lead to different scenarios of sulfur-polymer matrix composite and Portland cement composite service life.

ACKNOWLEDGMENTS

The described research was carried out as a part of the project funded by the Ministry of Education, Science and Technological Development of the Republic of Serbia under the Project Number 33007. The authors would like to extend their appreciation to the Ministry for this support.

REFERENCES

[1] Vlahović, M; Jovanić, P; Martinović, S; Boljanac, T; Volkov-Husović, T. Influence of Chemical Stress on Sulfur-Polymer Composite Structure, Chapter 10 in: *New Developments in Polymer Composites Research* (Editors: Stephan Laske and Andreas Witschnigg) Nova Science; Publishers, Inc., New York, United States of America, (2013).

[2] Vlahović, M. Doctoral dissertation. University of Belgrade, Faculty of Technology and Metallurgy, 2012.

[3] http://www.shell.com.

[4] Gracia, V; Vàzquez, E; Cramona, S. Utilization of by-produced sulfur for the manufacture of unmodified sulfur concrete, *International RILEM Conference on the Use of Recycled Materials in Buildings and Structures, Barcelona*, Spain, 2004, 1054-1063.

[5] World Business Council for Sustainable Development report, *Cement Industry Energy and CO_2 Performance, Getting the Numbers Right*, 2009.

[6] Vlahovic, M; Martinovic, S; Boljanac, T; Jovanic, P; Volkov-Husovic, T. Durability of sulfur concrete in various aggressive environments, *Construction and Building Materials*, 25, 3926-3934 (2011).

[7] *Use of sulfur concrete in precast applications*, Document Prepared by Leading Carbon Ltd., 2011, http://v-c-s.org/sites/ v-c-s.org/files/ Shell %20Concrete%20Methodology_for%20public %20comment. pdf.

[8] Mohamed, AMO; El Gamal, MM. *Sulfur concrete for the construction industry, A Sustainable Development Approach*, J. Ross Publishing, USA (2010).

[9] Diehl, L. Dyciclopentadiene-modified sulfur and its use as a binder, quoting sulfur concrete as an example, *New Uses for sulfur and Pyrites*. The Sulfur Institute, Madrid, 1976, 202-214.

[10] Beaudoin, JJ; Feldmant, RF. Durability of porous systems impregnated with dyciclopentadiene-modified sulfur. *The*

International Journal of Cement Composites and Lightweight Concrete, 1984, 6(1), 13-18.

[11] Sullivan, TA; McBee, WC; Blue, DD. Sulfur in coatings and structural materials in: West J.R. (Ed.), *New uses of sulfur: Advances in Chemistry Series 140*. American Chemical Society, Washington, D.C., 1975, 55-74.

[12] Gregor, R; Hackl, A. A new approach to sulfur concrete, in: Bourne D.J. (Ed.), *New uses of Sulfur-II: Advances in Chemistry Series 165*. American Chemical Society, Washington, D.C., 1978, 54-78.

[13] Lin, SL; Lai, JS; Chian, ESK. Modification of sulfur polymer cement (SPC) stabilization and solidification (S/S) process. *Waste Management*, 1995, 15, 441-7.

[14] Gannon, CR; Wombles, RH; Hettinger, WP; Watkins, WD. *New concepts and discoveries related to the strength of plasticized sulfur*. STP 807, ASTM, Philadelphia, 1983, 84-101.

[15] Mohamed, AMO; El Gamal, MM. Hydro-mechanical behavior of a newly developed sulfur polymer concrete. *Cement and Concrete Composites*, 2009, 31, 186-94.

[16] *Guide for Mixing and Placing Sulfur Concrete in Construction, Report* 548.2R-93. ACI Committee 548 (American Concrete Institute), Farmington Hills, Mi (1993).

[17] Dotto, JMR; De Abreu, AG; Dal Molin, DCC; Műller, IL. Influence of silica fume addition on concretes physical properties and on corrosion behaviour of reinforcement bars. *Cement and Concrete Composites*, 2004, 26, 31-39.

[18] Moosberg-Bustnes, H; Lagerblad, HB; Forssberg, EE. The function of fillers in concrete. *Materials and Structures*, 2004, 37, 74-81.

[19] Abdel-Jawad, Y; Al-Qudah, M. The combined effect of water and temperature on the strength of sulfur concrete. *Cement and Concrete Research*, 1994, 24, 165-75.

[20] Mohamed, AMO; El Gamal, MM. Sulfur bazed hazardous waste solidification, *Environmental Geology*, 53(1) 2012, 159-175.

[21] Mohamed, AMO; El Gamal, MM. Solidification of cement kiln dust using sulfur binder, *Journal of Hazardous Materials*, 192 (2) 2011, Pages 576-584.

[22] López, FA; Gázquez, M; Alguacil, FJ; Bolívar, JP; García-Díaz, I; López-Coto, I. Microencapsulation of phosphogypsum into a sulfur polymer matrix: physico-chemical and radiological characterization, *Journal of Hazardous Materials*, 192(1), 2011, 234-245.

[23] Adams, JW; Bowerman, BS; Kalb, PD. *Sulfur Polymer Stabilization/Solidification (SPSS) Treatability of Simulated Mixed-Waste Mercury Contaminated Sludge,* BNL-52648 Formal Report, 2001.

[24] Książek, M. The experimental and innovative research on usability of sulphur polymer composite for corrosion protection of reinforcing steel and concrete. *Composites Part B, Engineering*, 2011, Vol. 42, No. 5, pp. 1084-1096.

[25] Vlahović, M; Martinović, S; Jovanić, P; Boljanac, T; Volkov-Husović, T. Image Analysis Technique for Evaluating Damage Evolution and Predicting Mechanical Strength of Concrete Structures under Corrosion, Chapter 7 in: *Advances in Image Analysis Research* (Ed. Roger M. Echon), 2014, Nova Science Publishers, Inc. New York, Computer Science, Technology and Applications.

[26] Vidojković, V; Đorđević, N; Boljanac, T; Vlahović, M; Martinović, S; Branković, A. Procedure for cooling molten sulphur modified by dicyclopenthadiene on a flat metal surface. *Serbian Patent, 53412* (2014).

[27] Manufacture of sulfur concrete, *U. S. Patent 4025352* (1977).

[28] Concrete formulation comprising polymeric reaction product of sulfur/ cyclopentadiene oligomer/ dyclopentadiene, *U. S. Patent, 4348313* (1982).

[29] Acid resistant concrete articles, especially sulfur concrete pipes, and a method of manufacturing said articles, *DK Patent, 4981740* (1991).

[30] Drop forming method and sulphur polymer concrete process, *U. S. Patent, 6228905* (2001).

[31] Vlahović, M; Boljanac, T; Branković, A; Vidojković, V; Martinović, S; Đorđević, N. The influence of filler type on the corrosion stability of the sulfur concrete, *Hemijska industrija*, 64 (2), 129-137, (2010).

[32] Vlahović, M; Jovanić, P. Scanning electron microscopy analysis of sulfur-polymer composite subjected to induced destruction *Microscopy Book Series- Number 5: Current microscopy contributions to advances in science and technology,* Volume 2, Ed. Antonio Méndez-Vilas, 2012, 1171-1182, Formatex Research Center, Spain.

[33] Vlahović, M; Jovanić, P; Martinović, S; Boljanac, T; Volkov-Husović, T. Quantitative evaluation of sulfur-polymer matrix composite quality, *Composites Part B: Engineering*, DOI: 10.1016/j.compositesb.2012.04.005 (2012).

[34] *Image Pro Plus, Version 6.2 for Windows*, Media Cybernetics, Silver Spring.

[35] Russ, JC. *The Image Processing Handbook*, Second edition, CRC Press, (1995).

[36] Jovanic, P. Quantification of Visual Information. In: Hsu JP, Spasić AM, eds. *Finely dispersed particles: Micro-, nano-, and atto-engineering*. CRC/Taylor and Francis Group, Boca Raton, 341-364 (2005).

[37] Vlahović, M; Martinović, S; Boljanac, T; Majstorović, J; Volkov Husović, T. Application of ultrasonic measurements on durability testing of sulfur concrete, *9th Scientific/Research Symposium with International Participation "Metallic and Nonmetallic Materials", Zenica, B&H, Proceedings*, 311-316 (2012).

[38] Vlahović, M; Martinović, S; Majstorović, J; Volkov Husović, T. Assessment of sulfur-polymer composite quality based on nondestructive testing; Invited lecture, *The 46th International October Conference on Mining and Metallurgy,* 01- 04 October 2014, Bor Lake, Bor (Serbia), Proceedings, pages 40-43.

[39] Vlahović, M; Martinović, S; Volkov Husović, T. Valorization of secondary sulfur from oil refining process for sulfur concrete production, Plenary lecture, *XXIII International Conference*

"Ecological Truth" Eco -Ist' 15, Kopaonik, Serbia 17-20 June 2015, Proceedings, 12-28.

[40] Vlahović, M; Savić, M; Martinović, S; Boljanac, T; Volkov-Husović, T. Use of image analysis for durability testing of sulfur concrete and Portland cement concrete, *Materials and Design*, 34, 346-354 (2012).

[41] Vlahović, M; Jovanić, P; Martinović, S; Volkov Husović, T. Behavior of sulfur- polymer matrix composites with different fillers under acid and salt influence, Chapter 9 in *Polymer-Matrix Composites: Types, Applications and Performance*, Ed. Rakesh Kumar, Publisher: Nova Science Publishers, Inc., New York, United States of America, (2014), 215-240.

[42] Vlahovic, M; Martinovic, S; Boljanac, T; Jovanic, P; Volkov Husovic, T. Assesment of sulfur concrete corrosion resistance, *43rd International October Conference on Mining and Metallurgy, Serbia, Proceedings*, 370-373 (2011).

[43] Vlahović, M; Vidojković, V; Martinović, S; Boljanac, T; Jančić Heinemann, R; Volkov Husović, T. Corrosion resistance of sulfur concrete in acidic solution, *42nd International October Conference on Mining and Metallurgy, Serbia, Proceedings*, 102-105 (2010).

[44] Vlahovic, M; Martinovic, S; Boljanac, T; Volkov Husovic, T. Strength degradation modeling based on non-destructive testing, *43rd International October Conference on Mining and Metallurgy, Serbia, Proceedings*, 366-369 (2011).

In: Polymer-Matrix Composites ISBN: 978-1-53613-398-1
Editors: Elias Rice and Beth Sparks

Chapter 2

AN INVESTIGATION OF ELECTRON BEAM IRRADIATION ON ENHANCING COPPER (I) OXIDE IN A POLYETHYLENE MATRIX COMPOSITE

Soo-Tueen Bee[1,*], Lee Tin Sin[1,†], C. T. Ratnam[2], Fu-Yen Au[1] and A. R. Rahmat[3]

[1]Department of Chemical Engineering, Lee Kong Chian Faculty of Engineering and Science, Universiti Tunku Abdul Rahman, Jalan Sungai Long, Bandar Sungai Long, Cheras, Selangor, Malaysia
[2]Radiation Processing Technology Division, Malaysian Nuclear Agency, Bangi, Kajang, Selangor, Malaysia
[3]Department of Polymer Engineering, Faculty of Chemical Engineering, Universiti Teknologi Malaysia, UTM Skudai, Johor, Malaysia

[*] Corresponding Author Email: Soo-Tueen Bee: direct.beest@gmail.com.
[†] Corresponding Author Email: Lee Tin Sin: direct.tinsin@gmail.com.

ABSTRACT

In this work, the effects of electron beam irradiation (EB) on copper (I) oxide added polymer matrix composite (PMC) was investigated. Low density polyethylene (LDPE) was blended with copper (I) oxide using an extruder while further undergone EB at 50-300 kGy to induce crosslinking in the PMC. Such copper (I) oxide added PMC has been known for engineering applications that involving plastics moving parts to reduce friction. It was found that the EB has interacted with copper (I) oxide to form crosslinking network in the PMC matrix for stronger structures as exhibited by the gel content of PMC were higher than pristine LDPE at higher irradiation dosages (≥100 kGy). The formation of crosslinking networks in polymer matrix of PMC could resist the structural reorganization of polymer matrix when subjected to straining stress. The occurrence of fibrils on the facture surface of the PMC proved that EB crosslinking enables evenly load transfer when subjected to straining as well. Finally, the application of irradiation dosage up to 100 kGy and 200 kGy also significantly increased the crystallinity of PMC by the formation of crosslinking networks. The formation of crosslinking networks in LDPE matrix could rearrange the random chains structures into highly ordered chains arrangement. This is due to the EB irradiation could induce the formation of new crystallites in amorphous regions of the PMC.

Keywords: polymer-matrix composites (PMCs), mechanical properties, electron beam irradiation, crosslinking

INTRODUCTION

Since decades ago, there are large quantum increases of polymers applications involving sliding motion [1]. It was found that the introduction of additives especially the rigid fillers and reinforcements can affect the wear behaviour of polymer compound. According to Bahadur and Schwartz [3,4], wide range of inorganic compounds such as MoS_2, graphite, CuO, CuS, PbO, Fe_2O_3 and metallic powders have used as fillers in polymer to reduce wear corresponding to the reducing of coefficient of friction. In fact, according to Bahadur and Tabor [1] the addition of CuO, CuS, Pb_3O_4 and graphite can reduce the wear of high density polyethylene for about 20%. For polytetrafluoroethylene (PTFE), the rate of wear has significantly reduced with the incorporation of zinc, nickel, selenium, Y_2O_3 and SiO_2 [5]. Generally, the reduction of wear by the addition of metal filler is due to the modification of counterface surface and also because fiber supported of the applied load [6]. Bahadur and Gong [7] explained that the metal filler typically CuO can form microfragments to show roll-like structure than reduce the wear instantaneously. Nevertheless, the addition of incompatible metal filler may lead to deterioration of the polymer composite [8]. This is especially important in the wear engineering application that alternatives need to be taken such as through implementation of crosslinking structure in order to further enhance its ability to withstand harsh applications.

Radiation crosslinking is one of the methods used to improving the properties of polymer by introducing the formation of crosslinking networks in polymer matrix [9-11]. Recently, the importance of an irradiation technique has been significantly increased as an alternative method to induce the crosslinking networks in polymers matrix besides application of chemical crosslinking agent. This is due to the application of irradiation technique could eliminate the remaining chemical residues from chemical crosslinking agents in polymer which might be harmful to the consumers. Generally, there are various forms of high energy irradiations such as electron beam irradiation, gamma rays irradiation and X-ray irradiation which are currently used to modify the properties of polymer materials [12]. The application of irradiation technique on polymers can induce the formation of crosslinking by generating polymeric free radicals in polymer matrix. These free radicals could crosslink two long macromolecule chains by forming C-C intermolecular bonds to modify the properties of polymers. In general, two important mechanisms such as crosslinking process and chain scissioning reaction would be occurred during the irradiation process of polymers. The dominant mechanism could affect the changes in matrix structure and also mechanical properties of polymers [13-14]. The mechanical-structural properties of polymer matrix could be significantly enhanced when the crosslinking reaction is more dominance than chain scissioning reaction during the irradiation process.

This is due to the formation of crosslinked networks in polymer matrix could significantly increase the molecular weights of polymers [15]. On the other hand, the chain scissioning reaction is pre-dominated during the irradiation process, the polymer matrix could be severely degraded by breaking down the macromolecules in polymer into smaller sizes. Thus, this could further reduce the molecular weight of polymer and weaken the mechanical-structural properties of polymer [15]. In order to achieve predomination of crosslinking reaction compared to chain scissioning reaction during irradiation process, the application of suitable irradiation dosage is important for optimising the release of high energy free radicals to achieve the effective crosslinking.

According to previous studies [11, 16], LDPE tends to crosslink and form a three dimensional polymer network structure when exposed to high energy electron beam irradiation. The application of electron beam irradiation dosage up to 300 kGy on LDPE could induce the formation of crosslinking networks in LDPE matrix without undergoing degradation. LDPE polymer also can be easily crosslinked without the addition of any crosslinking agents due to its higher branched structure [17]. The high radiation resistance ability of LDPE has caused the electron beam irradiation technique to be used in accomplishing the improvement in LDPE properties. Generally, the formation of crosslinking networks in LDPE matrix by high energy irradiation involved two main mechanism stages [18]. The initial stage of crosslinking mechanism takes place by breaking down of C-H bond on LDPE chains to form free radicals by releasing hydrogen atom. Then, the released hydrogen atom would react with hydrogen atom released by other LDPE chains to form hydrogen gas. The generated free radicals from initial stage would react and combine together to form crosslinking network during the second stage of crosslinking mechanism [18]. In short, the electron beam irradiation of LDPE could easily generate free radicals and the generated free radicals could form three dimensional networks in LDPE matrix by releasing hydrogen gas. The increasing of irradiation dosage could rapidly induce the degree of crosslinking in LDPE matrix. Such crosslinked LDPE was found to exhibit outstanding mechanical properties such as tensile strength, elongation, etc. and electrical properties [16, 19-20]. In this study, polymer-matrix-composite (PMC) was prepared from blending of copper (I) oxide with low density polyethylene (LDPE) to investigate its reactivity when undergone electron beam irradiation. The outcomes of this work are important to provide an alternative to PMC in engineering field involving moving parts applications.

EXPERIMENTAL

Materials

Low density polyethylene (LDPE) with commercial grade of NM 1209 P3Y was manufactured by Petlin (Malaysia) Sdn. Bhd. LDPE was used as polymer matrix in this study. Copper (I) oxide with analytical reagent (AR) grade was supplied by ChemSolutions Sdn. Bhd. Copper (I) oxide was used as filler in this study.

Samples Preparation

LDPE were compounded with 1-5 phr (part per hundred resin polyethylene) of copper (I) oxide respectively using Brabender mixer at the heating temperature of 135°C for a mixing time of 8 minutes. During the compounding process, the rotor speed of Brabender mixer used is 50 rpm. The compounded samples were further hot pressed into 1 mm thickness sheet using hot press machine under the compression pressure of 10 MPa for 10 minutes. Initially, the compounded samples were preheated at heating temperature of 185°C for 5 minutes in hot press machine. Then, the preheated samples were further compression moulded at heating temperature of 185°C and pressure of 10 MPa for 5 minutes. The compression moulded samples were then further cooled down to room temperature under the pressure of 10 MPa for 2 minutes by using cold press with cooling rate of 15°C/min. The 1 mm thickness sheets were cut into required shapes using suitable sample cutter. The cut samples were then electron beam irradiated at the room temperature for 50-300 kGy. The electron beam irradiation rate used is 50 kGy per pass.

Gel Content

The gel content test was conducted by immersing the samples gravimetrically in hot xylene in accordance to ASTM D2765. The gel content test was carried out to determine the degree of crosslinking networks formed in polymer matrix. Firstly, the samples were accurately weighed using analytical balance to obtain initial weight (W_i). The weighted samples were immersed and heated in hot xylene at the temperature of 120°C for 24 hours. After the extraction process, the remaining samples were washed and rinsed with clean xylene for several times to remove the stain of soluble materials on the extracted samples. After that, these samples were rinsed again with methanol twice and soaked in methanol for 20 minutes. The washed samples were then dried in a vacuum oven to constant weight at 60°C for six hours. The weight of dried samples or insoluble materials (W_f) was obtained using analytical balance. The gel content percentage of the sample was calculated as shown in equation (1). The final value of gel content percentage for each sample was taken from the average of three specimens.

$$Gel\ content\ \ percentage\ (\%) = \frac{W_f}{W_i} \times 100\% \quad (1)$$

where W_i = Initial weight of the samples before the extraction process (g) and W_f = Final weight of the dried remaining samples after the extraction process (g).

Hot Set Test

The hot set test was carried out in a hot set oven (Heraeus UT 6050 HS) from Germany in accordance to standard of DIN 57472 Part 602 VDE 0472. The hot set test was performed to investigate the degree of crosslinking in polymer matrix of samples by determining the creep elongation of samples under a static load and heating temperature of 200°C. The 1 mm thickness sheet was cut into dumbbell shape specimens (similar to the specimen shape for tensile test). The initial gauge length was marked on the dumbbell specimens. After 15 minutes of heating, the elongation between the two marks on the dumbbell specimens was measured. The final elongation value of samples was recorded as an average of three specimens.

Tensile Test

The tensile test was carried out to investigate the tensile properties of samples under straining stress in according to ASTM D1822. The 1 mm thickness sheets were cut into dumbbell shape using dumbbell cutter. The dumbbell specimens were tested by using Instron Universal Testing Machine with model 4302 Series IX. The gauge length of all the dumbbell specimens was fixed at 10 mm. The gauge thickness of samples was measured before the tensile testing. The tensile test was carried out at room temperature of 25°C under load of 2 kN with the crosshead speed of 50 mm/min. The tensile strength and elongation at break were taken as an average of six specimens.

Morphology Observation

The surface morphologies of the fractured surface for the copper (I) oxide added LDPE composites from tensile test were scanned and observed using Hitachi Scanning Electronic Microscopy (SEM) of BS 340

TESLA. The fractured surfaces of the samples were cut into a smaller portion before the SEM scanning. Then, the cut samples were placed and mounted onto the copper stub with the fractured surface was facing up. The mounted samples were coated with a thin layer of gold and ready for scanning. The SEM micrographs were recorded at a magnification of 1000 times at the electron beam voltage of 15 kV.

X-Ray Diffraction (XRD) Test

The crystallinity and crystalline structure of all copper (I) oxide added LDPE composites was evaluated by using X-ray Diffraction (XRD). The XRD spectra of all LDPE samples was scanned by using Shimadzu XRD 6000 diffractometer with Cu-Kα radiation (λ = 1.542 Å) for a 2θ range of 0° to 40° at a scanning rate of 1.2°/min. The Cu-Kα radiation generator was operated at accelerating voltage and current of 30 kV and 30 mA, respectively. The degree of crystallinity was calculated as the percentage of the scattered intensity of the crystalline phase over the scattered intensity of the crystalline and amorphous phase. The crystalline size, *L* was derived from the peak angle (2θ) in the XRD diffractogram in accordance to the Scherrer equation as shown in equation (2).

$$L = \frac{k\lambda}{\beta \cos\theta} \quad (2)$$

where *k* is Scherrer constant (usually taken as 1.0), β is the full width at half maxima (FWHM) of the diffraction peak, λ is wavelength of Cu-irradiation (λ = 1.542 Å) and 2θ is the angle of diffraction peak in radians. The d-spacing of crystallite structure were calculated using the Bragg's equation as shown in equation (3). On the other hand, the inter-chain separation, *R* of crystalline structure was calculated using Klug and Alexander equation as shown in equation (4).

$$d = \frac{\lambda}{2\sin\theta} \tag{3}$$

$$R = \frac{5\lambda}{8\sin\theta} \tag{4}$$

where 2θ is the angle of diffraction peak in radians and λ is 1.542 Å.

RESULTS AND DISCUSSION

Gel Content Analysis

The gel content test was carried out to evaluate the degree of crosslinking formed in polymer matrix. All the un-irradiated copper (I) oxide added LDPE composites were found to fully dissolve in hot xylene after 24 hours of extraction as shown in Figure 1. This is due to the absence of crosslinking networks in LDPE matrix has caused the LDPE matrix unable to resist the attack of hot xylene [16]. By referring to Figure 1, the gel content of all the LDPE composites was observed to rapidly increase when irradiated up to 50 kGy. The application of electron beam irradiation on LDPE matrix could generate polymeric free radicals in LDPE matrix by attacking the -C-H bonding of LDPE chains in amorphous regions to release hydrogen ion as shown in reaction 5 [12, 16]. These released hydrogen ions could further extract the hydrogen atoms from the other LDPE chains to generate the new polymeric free radicals by forming gas hydrogen as shown in reaction 6. The polymeric free radicals formed could react together to form a crosslinking network in between two LDPE chains as shown in reaction 7 [12]. The formation of crosslinking networks in LDPE matrix (as shown in reaction 7) could highly increase the resistance of LDPE matrix towards the attack of hot xylene by reducing the solubility of LDPE matrix in hot xylene [20]. Thus, the gel content of all copper (I) oxide added LDPE composites was rapidly increased when irradiated to 50 kGy.

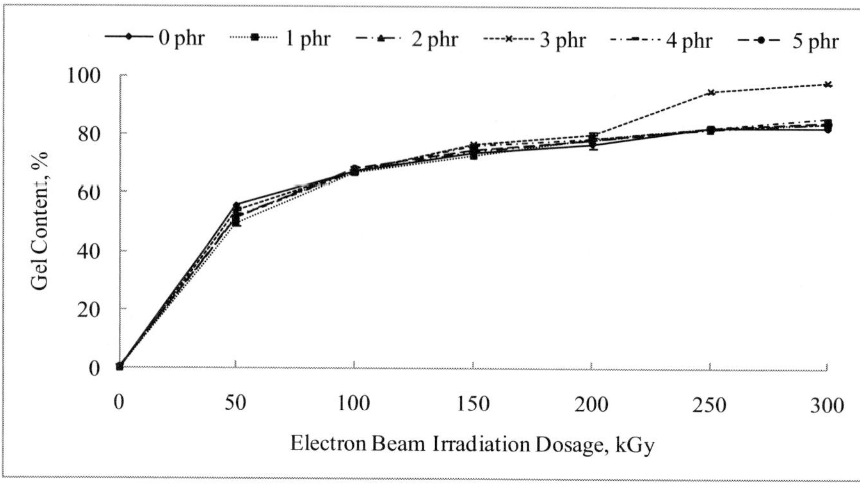

Figure 1. Gel content of LDPE composites added with increasing of copper (I) oxide loading level when subjected to various electron beam irradiation.

$$-(CH_2-CH_2)_n- \xrightarrow{electron} -(CH_2-C^*-H)_n- \; + H^+ \quad (5)$$

$$-(CH_2-CH_2)_n- + H^+ \rightarrow -(CH_2-C^*-H)_n- \; + H_2(gas) \quad (6)$$

$$\begin{matrix} -(CH_2-CH^*)_n- & & -(CH_2-CH)_n- \\ & \rightarrow & | \\ -(CH_2-CH^*)_n- & & -(CH_2-CH)_n- \end{matrix} \quad \text{(crosslinked chains)} \quad (7)$$

By referring to Figure 1, further increment in irradiation dosages from 50 kGy to 300 kGy has marginally increased the gel content of all copper (I) oxide added LDPE composites. This also indicates the increasing of gel content (or also known as the degree of crosslinking networks formation) in LDPE matrix is strongly dependent to the level of irradiation dosages. The increasing of gel content value indicates the higher portion of amorphous region in LDPE matrix formed with crosslinking networks [16]. However, the rates of gel formation for all copper (I) oxide added LDPE composites were also observed to be significantly reduce at higher irradiation dosages (≥100 kGy). At higher irradiation dosages, the already crosslinked chains in LDPE matrix have restricted the mobility of polymeric free radicals generated in LDPE matrix by entrapping these free radicals inside the crosslinking networks. These trapped polymeric free radicals are likely to promote the formation of chemical bonding inside the already crosslinked chains. The formation of these crosslinking bonds in between the already crosslinked chains would not significantly contribute to the increment in degree of crosslinking networks (gel content) of all copper (I) oxide added LDPE composites [20].

At lower irradiation dosage (50 kGy), all the copper (I) oxide added LDPE composites were observed to pose lower gel content values than pristine LDPE. This might be due to the occurrence of copper (I) oxide particles in LDPE matrix could act as blockage to hinder the formation of crosslinking networks in LDPE matrix [22]. However, it is interesting to note that the increasing of copper (I) oxide loading level from 1 phr to 3 phr has marginally increased the gel content of 50 kGy irradiated LDPE composites from 49% to 54%. Under electron beam irradiation, the copper (I) oxide in LDPE matrix could absorb two accelerated electrons (from electron beam accelerator) to form copper and ion O^{2-} as shown in reaction 8 [8]. The generated ion O^{2-} could extract two hydrogen atoms from LDPE chains to form water by releasing the polymeric free radicals as illustrated in mechanism reaction 9 [8]. This mechanism reaction could slightly induce the amounts of polymeric free radicals generated in LDPE matrix and thus contribute to the higher degree of crosslinking networks formed in LDPE matrix. However, further increasing the copper (I) oxide loading level from 3 phr to 5 phr did not pose significant effect in promoting the formation of crosslinking networks. This might be due to higher content of copper (I) oxide tended to generate the polymeric free radicals through chain scissioning of LDPE chains [23]. Thus, this would not contribute any significant effect on the degree of crosslinking (gel content) in LDPE matrix.

$$Cu_2O \;+\; 2\bar{e} \;\rightarrow\; 2\,Cu \;+\; O^{2-} \qquad (8)$$

$$2\ -(CH_2 - CH_2)- \ + \ O^{2-} \ \rightarrow \ 2\ -(CH_2 - C^{*}H)- \ + \ H_2 0 \tag{9}$$

However, the gel content of all copper (I) oxide added LDPE composites were observed to be slightly higher than gel content result of pristine LDPE at higher irradiation dosages (≥100 kGy) as shown in Figure 1. This might be due to the addition of copper (I) oxide in LDPE matrix could slightly induce the formation of polymeric free radicals under electron beam irradiation through the mechanism reactions 8 and 9 as discussed earlier [8]. At the irradiation dosages increased from 200 kGy to 300 kGy, all copper (I) oxide added LDPE composites show an insignificant increment in gel content results except for the 3 phr copper (I) oxide added LDPE composite. The application of higher electron beam irradiation dosages up to 250 kGy and 300 kGy on 3 phr copper (I) oxide added LDPE composite could simultaneously generate more effective polymeric free radicals in LDPE matrix for crosslinking. Thus, this could further induce the effectiveness in promoting the formation of crosslinking networks in polymer matrix of 3 phr copper (I) oxide added LDPE composite.

Hot Set Test

The hot set results for all un-irradiated and irradiated copper (I) oxide added LDPE composites under heating temperature of 200°C and subjected to static load of 10 N/cm^2 were presented in Table 1. By referring to Table 1, all the un-irradiated samples were observed to break immediately when subjected to heating temperature of 200°C and static load of 10 N/cm^2. This is due to the un-irradiated LDPE matrix unable to withstand the static load under high temperature without the occurrence of crosslinking networks. The absence of crosslinking networks in LDPE matrix caused the macromolecules chains in LDPE matrix can easily move and slippage on each other without restriction when subjected to static load under high temperature [14, 19]. This observation is in agreement with the gel content measurement results, where no gel content (degree of crosslinking) was detected in the polymer matrix of all un-irradiated samples.

All the copper (I) oxide added LDPE composites remained to fail within 15 minutes even electron beam irradiated to 50 kGy. However, the failing time of all 50 kGy irradiated- copper (I) oxide added LDPE composites was observed to slightly delay by 1 to 2 minutes in comparing to all un-irradiated samples. When the irradiation dosages increased from 50 kGy to 200 kGy, all the samples still remained falling within 15 minutes under high temperature. The failure time of all copper (I) oxide added LDPE composites was gradually delayed up 7 to 10 minutes. The degree of crosslinking networks in LDPE matrix could be effectively induced by electron beam irradiation by generating more free radicals in LDPE matrix [24]. The formation of higher degree of crosslinking networks in LDPE matrix could significantly prolong the resistance time of LDPE matrix towards the static load under high temperature. The failure time of 3 phr copper (I) oxide added LDPE composite was found to be delayed up to 11 to 14 minutes when irradiated to 200 kGy which this failure time is found to be longer than all the other copper (I) oxide added LDPE samples. This also indicates the higher degree of crosslinking networks in polymer matrix of 3 phr copper (I) oxide added LDPE composite which it could help in improving the resistance effect of LDPE matrix at high temperature.

As the electron beam irradiation further increased from 200 kGy to 300 kGy, the creep elongation of all copper (I) oxide added LDPE composites was slightly decreased as shown in Table 1 and Figure 2. The increasing of electron beam irradiation dosages could significantly induce the generating of free radicals in LDPE matrix to form higher degree of crosslinking networks in LDPE matrix [16]. The formation of crosslinking networks in LDPE matrix could restrict the slippage effect of LDPE macromolecule chains in LDPE matrix when subjected to static load [19-21]. Thus, the elongation ability of LDPE matrix of all samples has been reduced with increasing of electron beam irradiation dosages from 250 kGy to 300 Gy. By referring to Figure 2, the increasing of copper (I) oxide loading level from 0 phr to 3 phr has gradually decreased the creep elongation of LDPE composites from 91% to 71%. This could be due to the copper (I) oxide particles in LDPE matrix could slightly induce the release of free radicals in LDPE matrix when subjected to electron beam irradiation in accordance to reaction mechanism as shown in equations 8 and 9 [8]. The higher free radicals amounts generated could slightly induce the degree of crosslinking networks in copper (I) oxide added LDPE composites. By referring to Figure 2, the creep elongation of 250 kGy and 300 kGy irradiated 3 phr copper (I) oxide added LDPE composite has reached the lowest values in comparing to the other copper (I) oxide added LDPE composites. This is due to the application of electron beam irradiation (250 kGy and 300 kGy) on 3 phr copper (I) oxide added LDPE composites could generate more effective free radicals for crosslinking simultaneously. However, the creep elongation of irradiated LDPE composites was gradually increased when the loading level of copper (I) oxide further increased from 3 phr to 5 phr. At higher loading level of copper (I) oxide (≥4 phr), the copper (I) oxide in LDPE matrix would attack the main backbone of LDPE chains to generate the free radicals in LDPE matrix, where this process tends to promote the occurrence of chain scissioning process instead of crosslinking [8].

Table 1. Hot set results of LDPE composites added with increasing loading level of copper (I) oxide when subjected to various electron beam irradiation dosages

Irradiation	Loading level of Copper (I) Oxide, phr					
dose, kGy	0	1	2	3	4	5
0	Fail immediately	Fail immediately	Fail immediately	Fail immediately	Fail immediately	Fail immediately
50	Fail at 1-2 min	Fail at 1-2 min	Fail at 1-2 min	Fail at 1-2 min	Fail at 1-2 min	Fail at 1-2 min
100	Fail at 3-5 min	Fail at 3-5 min	Fail at 3-5 min	Fail at 3-5 min	Fail at 3-5 min	Fail at 3-5 min
150	Fail at 5-8 min	Fail at 5-8 min	Fail at 5-8 min	Fail at 5-8 min	Fail at 5-8 min	Fail at 5-8 min
200	Fail at 5-10 min	Fail at 5-10 min	Fail at 5-10 min	Fail at 9-14 min	Fail at 5-10 min	Fail at 5-10 min
250	91 ± 2.5%	86 ± 2.5%	84 ± 2.1%	71 ± 2.5%	96 ± 3.5%	99 ± 3.1%
300	89 ± 3.0%	83 ± 2.6%	81 ± 3.0%	69 ± 3.1%	84 ± 2.6%	87 ± 2.1%

Mechanical Properties Analysis

By referring to Figure 3(a), the incorporation of copper (I) oxide particles into LDPE matrix has significantly decreased the tensile strength of un-irradiated and irradiated LDPE composites. This might be due to the occurrence of copper (I) oxide particles in LDPE matrix could cause disruption of LDPE structure arrangement by the dissimilar sphere metallic of copper (I) oxide particles [25]. According to Figure 3(a), the tensile strength of all LDPE composites was observed to severely decrease with the addition of 1 phr copper (I) oxide. As mentioned in previous, this might be due to the addition of copper (I) oxide particles could disturb the LDPE structure arrangement in polymer matrix. Besides, the addition of copper (I) oxide particles as filler in LDPE matrix also could reduce the intermolecular bonding between the LDPE macromolecules chains [26] and thus decreased the tensile strength of LDPE composites. However, the tensile strength was gradually increased when the loading level of loading level of copper (I) oxide increased from 1 phr to 3 phr. The higher loading level of copper (I) oxide in LDPE matrix could slightly reduce the irregularity arrangement of copper (I) oxide particles in LDPE matrix (inhomogeneous distribution) and improve the interfacial adhesion effect between the copper (I) oxide particles and LDPE matrix. The good interfacial adhesion effect between the copper (I) oxide particles and LDPE matrix could slightly induce the effectiveness in transferring the applied stress evenly throughout the polymer matrix [12]. Thus, the tensile strength has been enhanced. However, the tensile strength of all LDPE composites was observed to significantly reduce with further increasing the copper (I) oxide loading level from 3 phr to 4 phr and 5 phr as indicated in Figure 3(a). At higher loading level (≥4 phr) of copper (I) oxide, the agglomeration of copper (I) oxide particles are likely to occur in LDPE matrix due to the copper (I) oxide particles tended to adhere to each other instead of LDPE matrix. Thus, this could lead to poor interfacial adhesion effect between the agglomerated copper (I) oxide and LDPE matrix where the agglomerated particles could act as stress concentration points and unable to transfer the applied straining stress throughout the polymer matrix effectively.

The effect of electron beam irradiation on tensile strength of copper (I) oxide added LDPE composites is illustrated in Figure 3(a). By referring to Figure 3(a), the increasing of electron beam irradiation dosages from 0 kGy to 200 kGy has gradually increased the tensile strength of pristine LDPE. This observation is found to be consistent with the results of a research conducted by Basfar [27]. This is due to the application of electron beam irradiation could induce the formation of polymeric free radicals in LDPE matrix by attacking the C-H bonding of LDPE macromolecules chains to release gas hydrogen in accordance to the mechanism as shown in equation 5 and 6 [28]. The generated polymeric free radicals in LDPE matrix could react together to form crosslinking networks in LDPE matrix. The occurrence of crosslinking networks in LDPE matrix could further improve the tensile strength by providing the reinforcement effect to the LDPE matrix in resisting the applied straining stress. The increasing of electron beam irradiation dosages also could induce the formation of polymeric free radicals in LDPE matrix and this could further increase the degree of crosslinking networks formed in LDPE matrix. Thus, the tensile strength of pristine LDPE has been enhanced. However, further increasing the electron beam irradiation dosages from 200 kGy to 300 kGy has rapidly decreased the tensile strength of pristine LDPE. At higher irradiation dosages (>200 kGy), the available C-H bonding in LDPE macromolecule chains to generate polymeric free radicals in LDPE matrix has significantly reduced and tended to promote the occurrence of chain scissioning process. In this process, the accelerated electron (released by electron beam accelerator) would attack and breakdown the C-C bonding in backbone chains of LDPE macromolecules to generate polymeric free radicals in LDPE matrix [18]. The generated polymeric free radicals in LDPE matrix would form the peroxy species by reacting with oxygen [29]. The peroxy species formed would further decompose into smaller macromolecules and reduce the molecular weight of LDPE chains. Thus, the tensile strength has significantly reduced.

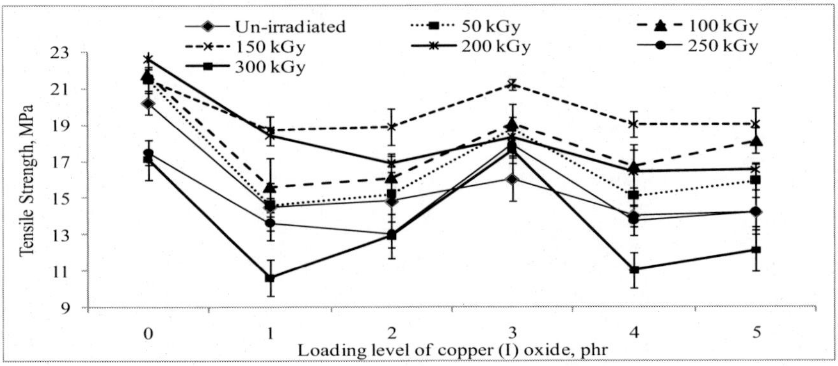

(a)

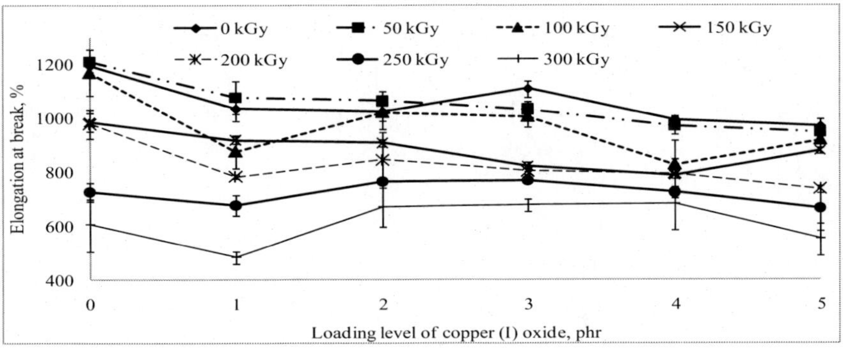

(b)

Figure 3. (a) Tensile strength and (b) elongation at break of LDPE composites added with increasing loading level of copper (I) oxide when subjected to various irradiation dosages.

By referring to Figure 3(a), the tensile strength of all copper (I) oxide added LDPE composites was observed to gradually increase when electron beam irradiated from 0 kGy to 150 kGy. The increment of electron beam irradiation dosages could induce the release of polymeric free radicals in LDPE matrix which could contribute to the formation of more crosslinking networks in LDPE matrix [19]. The formation of crosslinking networks in LDPE matrix could further improve the structural rearrangement of copper (I) oxide particles in LDPE matrix that caused by the irregularity shapes of copper (I) oxide particles. The formation of crosslinking networks in polymer matrix of copper (I) oxide added LDPE composites could prevent and resist the structural reorganization of polymer matrix when subjected to straining stress [20]. Thus, the tensile strength has been significantly enhanced by the introduction of electron beam irradiation. However, further increment of irradiation dosages from 150 kGy to 300 kGy has gradually reduced the tensile strength of all copper (I) oxide added LDPE composites as shown in Figure 3(a). This phenomenon might be due to the application of high irradiation dosages (≥200 kGy) on polymer matrix of all copper (I) oxide added LDPE composites could cause the occurrence of chain scissioning [20]. The occurrence of chains scissioning process in polymer matrix could reduce the density of crosslinking networks and caused an inferior effect to the tensile strength of all copper (I) oxide added LDPE composites.

The effect of copper (I) oxide loading level and electron beam irradiation dosages on the elongation at break of pristine LDPE and copper (I) oxide added LDPE composites is illustrated in Figure 3(b). By referring to Figure 3(b), the addition of copper (I) oxide particles in LDPE matrix was observed to significantly reduce the elongation at break of un-irradiated LDPE composites. This is due to the irregular shapes of copper (I) oxide particles could disrupt the structural arrangement of polymer chains in LDPE matrix which restricted the slippage effect of LDPE macromolecules chains in LDPE matrix when under straining stress [20]. On the other hand, the addition of copper (I) oxide particles into LDPE matrix also could significantly reduce the continuities of LDPE matrix and thus further reduce the plastic deformation ability of polymer matrix [29]. Besides, the poor compatibility between copper (I) oxide particles and LDPE matrix also posed a significant effect in reducing the extendability of LDPE matrix. The poor compatibility effect could cause the copper (I) oxide particles to act as stress concentration points in LDPE matrix which caused LDPE matrix unable to elongate easily before the breakdown of polymer matrix. As the loading level of copper (I) oxide increased from 1 phr to 2 phr, the effect on elongation at break of un-irradiated LDPE composites is insignificant. By referring to Figure 3(b), further increment in copper (I) oxide loading level from 2 phr to 3 phr has significantly increased the elongation at break of un-irradiated LDPE composites. This is due to the increasing of copper (I) oxide loading level up to 3 phr could reduce the incompatibility effect between the copper (I) oxide particles and LDPE matrix by fitting into the interfacial between copper (I) oxide particles and LDPE matrix. This could further improve the interfacial adhesion effect between copper (I) oxide particles and polymer matrix which the LDPE matrix able to resist the applied straining stress by extending the elongation ability polymer matrix [19, 30]. However, the further increasing of copper (I) oxide loading level was found to marginally decrease the elongation at break of un-irradiated LDPE composites. At higher loading level of copper (I) oxide (>3phr), the copper (I) oxide particles in LDPE matrix tended to agglomerate together to form larger copper (I) oxide aggregates which the occurrence of the copper (I) oxide agglomerated particles could weaken the interfacial adhesion effect with LDPE matrix. The poor interfacial adhesion effect between copper (I) oxide agglomerated particles and polymer matrix is attributed to irregular structure of copper (I) oxide agglomerated particles in LDPE matrix. This could further promote the incompatibility effect of copper (I) oxide particles with LDPE matrix and cause the agglomerated copper (I) oxide particles to act as stress concentration point in LDPE matrix when subjected to straining stress [31]. Thus, the elongation at break of un-irradiated LDPE composites was reduced at higher loading level of copper (I) oxide.

For pristine LDPE and LDPE composites added with low loading level of copper (I) oxide (≤2 phr), the application of electron beam irradiation dosages up to 50 kGy has marginally increased the elongation at break as shown in Figure 3(b). This indicates that the application of low irradiation dosage (≤50 kGy) on polymer matrix of copper (I) oxide added LDPE composites could extend the elongation ability of LDPE matrix under stretching. This could be attributed to the electron beam irradiation process could generate the polymeric free radicals in LDPE matrix to induce the formation of crosslinking networks in LDPE matrix. The formation of low crosslinking amounts in LDPE matrix could slightly extend the macromolecules size of LDPE chains and prolong the resistance time of

LDPE matrix when subjected to stretching stress [32]. However, the elongation at break of pristine LDPE and all copper (I) oxide added LDPE composites was found to gradually decrease with further increasing of irradiation dosages up to 300 kGy. At higher irradiation dosages (>100 kGy), the application of electron beam irradiation on polymer matrix tends to induce the release of polymeric free radicals inside the already crosslinked chains [14]. The mobility of polymeric free radicals released inside the already crosslinked chains have been restricted by entrapping inside the crosslinking networks in LDPE matrix. The trapped polymeric free radicals are more likely to form chemical bonding inside the already crosslinked chains [14]. The crosslinking bonds formed inside the already crosslinked chains could further restrict the movement ability (slippage effect) of LDPE macromolecule chains under stretching and thus reduced the extendability of LDPE matrix.

Morphologies Analysis

Figures 4 to 5 illustrate the fractured surface observations of LDPE composites added with increasing of copper (I) oxide loading level when subjected to various electron beam irradiation dosages. By comparing the Figures 4(a) to 4(d), the polymer matrix for pristine LDPE was found to be more continuous than all copper (I) oxide added LDPE composites with the appearance of rough flakes like structures. The occurrence of flakes like structures was attributed to the tearing effect of polymer matrix of pristine LDPE when subjected to the applied straining stress. Besides, the flakes like structures on fractured surface of pristine LDPE also found to be connected to the polymer matrix which indicating the higher continuous behaviour of pristine LDPE polymer matrix. The polymer matrix with higher matrix continuities could pose higher resistance effect towards the applied training stress, thus the tensile strength and elongation at break of pristine LDPE also observed to be the highest in compared to other copper (I) oxide added LDPE composites. The appearance of fibrils can be clearly seen on the fractured surface of LDPE composite added with 1 phr copper

(I) oxide as depicted in Figure 4(b). By comparing to Figure 4(a), the continuities of LDPE matrix has been significantly reduced by tearing the polymer matrix into smaller fibrils when adding with 1 phr copper (I) oxide. This is because the copper (I) oxide particles in LDPE matrix would act as stress concentration point under straining stress [19]. This could cause the straining stress applied on the copper (I) oxide added LDPE matrix unable to transfer evenly throughout the polymer matrix. Thus, the applied stress would focus on the copper (I) oxide particles in LDPE matrix and cause the polymer matrix surrounding the copper (I) oxide particles started to tear into fibrils with lower width diameter. Thus, the addition of copper (I) oxide in low loading level (1 phr) could slightly reduce the matrix continuities (Bee et al., 2013). This observation also found in agreement with the results of tensile strength results as discussed earlier where the addition of 1 phr copper (I) oxide into LDPE matrix could significantly reduce the tensile strength of LDPE.

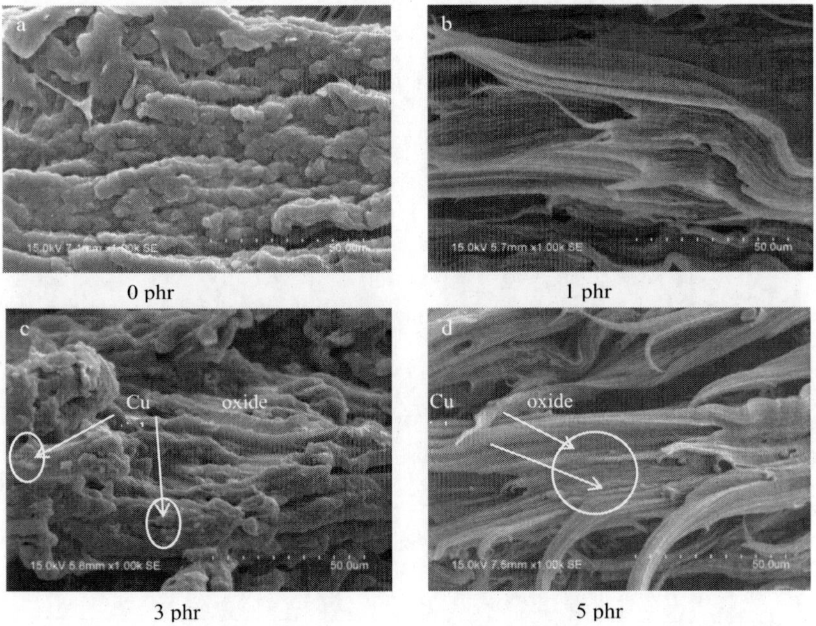

Figure 4. Surface morphologies at 1000x magnification of (a) pristine LDPE, LDPE composites added with (b) 1 phr copper (I) oxide, (c) 3 phr copper (I) oxide and (d) 5 phr copper (I) oxide.

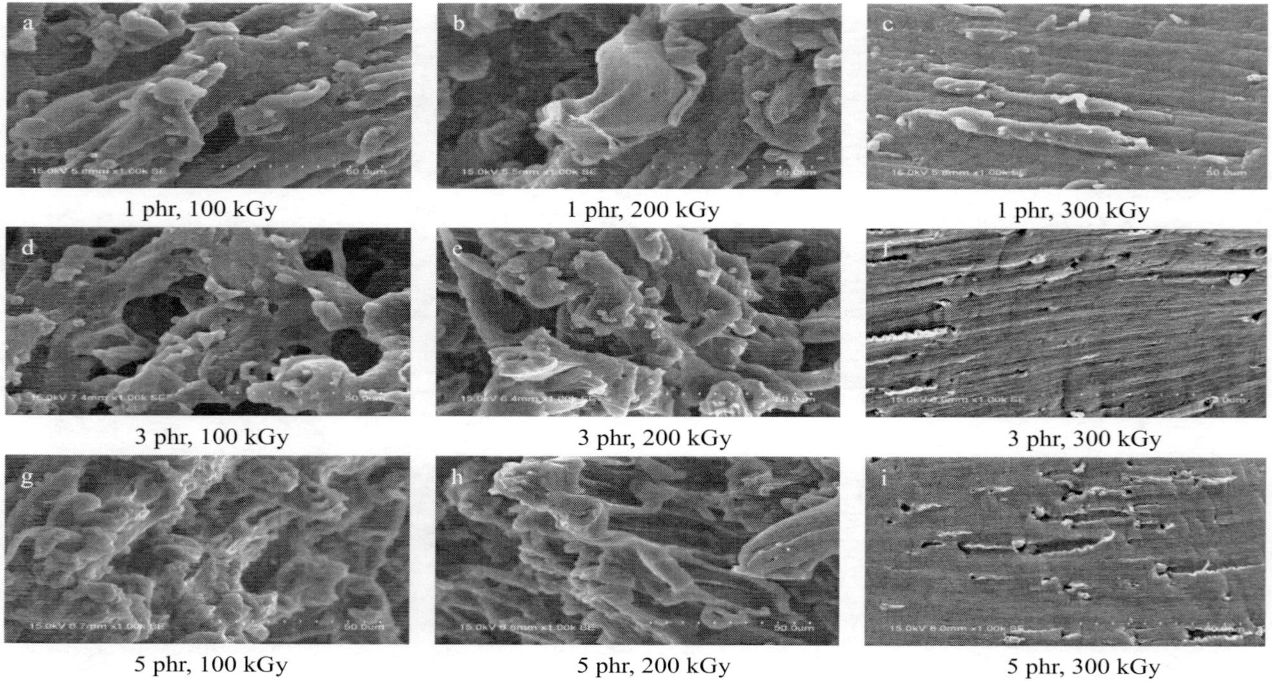

Figure 5. Surface morphologies of LDPE composites at 1000x magnification with 1 phr copper (I) oxide when subjected to (a) 100 kGy, (b) 200 kGy, (c) 300 kGy; LDPE composites added with 3 phr copper (I) oxide when subjected to (d) 100 kGy, (e) 200 kGy, (f) 300 kGy dosages and LDPE composites added with 5 phr copper (I) oxide under irradiation dosages of (g) 100 kGy, (h) 200 kGy and (i) 300 kGy.

By referring to Figure 4(c), the occurrence of fibrils also can be clearly observed on the LDPE matrix as the loading level of copper (I) oxide was increased up to 3 phr. However, it is interesting to note that the fibrils formed on fractured surface of 3 phr copper (I) oxide added LDPE matrix was observed to be much more larger in diameter width than the fibrils formed on the fractured surface of 1 phr copper (I) oxide added LDPE (as depicted in Figure 4(b)). This indicating the matrix continuities of LDPE have been increased as the loading level of copper (I) oxide increased from 1 phr to 3 phr. This could be due to the well dispersion and distribution of copper (I) oxide particles in LDPE matrix which could further enhance the interfacial adhesion effect (compatibility) between copper (I) oxide particles and LDPE matrix [19]. At higher loading level of copper (I) oxide (3 phr), the copper (I) oxide particles could interfacial adhesion effect between copper (I) oxide and LDPE matrix can be enhanced by modifying the structural arrangement of copper (I) oxide particles in LDPE matrix [20]. The good interfacial adhesion effect between copper (I) oxide particles and LDPE matrix could further improve the effectiveness in transferring the applied straining stress evenly throughout the whole polymer matrix. Thus, this could further reduce the number of copper (I) oxide particles to act as stress concentration point during straining which could promote the tearing of polymer matrix into thinner fibrils. However, further increasing in the loading level of copper (I) oxide from 3 phr to 5 phr was found to significantly reduce the matrix continuities of LDPE matrix as depicted in Figure 4(d). By referring to Figure 4(d), the occurrence of fibrils can be significantly seen on the fractured surface of 5 phr copper (I) oxide added LDPE composite. The fibrils formed was also observed to be thinner in diameter width and higher in amounts when compared to the fibrils formed on the fractured surface of 1 phr and 3 phr copper (I) oxide added LDPE composites as illustrated in Figure 4(b) and 4(c). Besides, the agglomeration of copper (I) oxide particles also can be clearly observed on the fractured surface of 5 phr copper (I) oxide added LDPE composite as shown in Figure 4(d). This is due to the poor dispersion effect of copper (I) oxide particles in LDPE matrix which cause the copper (I) oxide particles tended to agglomerate together into larger particles. The agglomeration of copper (I) oxide particles could severely reduce the interfacial adhesion effect between the agglomerated copper (I) oxide particles and LDPE matrix due to the irregular shapes of the agglomerated copper (I) oxide. This could cause the agglomerated copper (I) oxide particles with poor interfacial adhesion effect with LDPE matrix would act as stress concentration point when subjected to straining stress. This has caused the applied stress on polymer matrix unable to transfer effectively throughout the whole polymer matrix. Thus, the continuities of LDPE matrix has been significantly reduced by tearing LDPE matrix into larger number of fibrils with thinner in diameter width.

The effect of electron beam irradiation on surface morphologies of all copper (I) oxide added LDPE composites were depicted in Figure 5. By referring to Figures 5(a) to 5(b), the formation of fibrils on fractured surface of 1 phr copper (I) oxide added LDPE composite was observed to be disappeared as electron beam irradiated to 100 kGy and 200 kGy. On the other words, the polymer matrix of 100 kGy and 200 kGy irradiated 1 phr copper (I) oxide added LDPE composite was found to be more continuous than the un-irradiated samples. This is due to the application of electron beam irradiation could induce the formation of crosslinking networks in copper (I) oxide added LDPE matrix. The crosslinking networks formed in polymer matrix could increase extendability of LDPE matrix by increasing the macromolecular density of LDPE chains in LDPE matrix. Besides, the formation of crosslinking networks in LDPE matrix also could improve the interfacial adhesion effect between copper (I) oxide particles and LDPE matrix by providing a better structural rearrangement of copper (I) oxide particles in LDPE matrix. Thus, the matrix continuities of LDPE composite also found to be increased or extended. This observation is found to be consistent with the results of tensile strength (as discussed earlier) where the application of electron beam irradiation dosages up to 100 kGy and 200 kGy could gradually enhance the tensile strength of 1 phr copper (I) oxide added LDPE composite. However, the fracture surface of 1 phr copper (I) oxide added LDPE matrix was found to be rough and formed a brittle like structures when further electron beam irradiated up to 300 kGy as shown in Figure 5(c). The rough and brittle like structures of LDPE matrix indicating the elongation ability of LDPE matrix has been severely retarded by the application of high electron beam irradiation dosage (≥300 kGy). This is due to the application of high electron beam irradiation dosage up to 300 kGy has caused the chain scissioning process to be more pre-dominated. During the chain scissioning process, the backbone of LDPE macromolecule chains inside LDPE matrix would breakdown into smaller macromolecules chains [33]. This could further cause the polymer matrix unable to elongate well in resisting the applied straining stress before the breakdown of polymer matrix as illustrated in Figure 5(c). Thus, the matrix continuities of 300 kGy irradiated sample have been significantly reduced. Besides, the occurrence of small voids also can be observed on LDPE matrix which is attributed to the release of gas during chains scissioning process.

By referring to Figures 5(d) and 5(g), the application of electron beam irradiation dosage up to 100 kGy was observed to slightly increased the matrix continuities of 3 phr and 5 phr copper (I) oxide added LDPE matrix in comparing to un-irradiated samples (as shown in Figures 4(c) and 4(d)). This is attributed to the application of electron beam irradiation up to 100 kGy could induce the degree of crosslinking networks formed in LDPE matrix which the crosslinking networks formed could extend the polymer matrix continuities. Besides, the crosslinking networks formed also could improve the dispersion effect of copper (I) oxide particles in LDPE matrix by modifying the structural arrangement of copper (I) oxide particles in LDPE matrix. As the electron beam irradiation dosage increased up to 200 kGy, the surface morphologies of 3 phr and 5 phr copper (I) oxide added LDPE composites was found to be in similar pattern with the polymer matrix of 100 kGy irradiated samples. This also indicates that the degree of crosslinking formed in LDPE matrix posed insignificantly effect on the matrix continuities of 3 phr and 5 phr copper (I) oxide added LDPE composites when the electron beam irradiation dosages increased from 100 kGy to 200 kGy. However, further increment in electron beam irradiation up to 300 kGy has significantly reduced the matrix continuities of 3 phr and 5 phr copper (I) oxide added LDPE samples by forming the rough and brittle like structures on the polymer matrix as observed in Figures 5(f) and 5(i). Under higher electron beam irradiation dosages (≥300 kGy), the electron from electron beam accelerator tended to attack the backbone of LDPE chains to form polymeric free radicals and this reaction also known as chains scissioning process. The occurrence of chain scissioning process could significantly reduce the macromolecules sizes of LDPE chains (or density of macromolecular weight). This could rapidly reduce the extendability of LDPE matrix by reducing resistance ability of LDPE towards the applied straining stress and further reduce the matrix continuities of LDPE matrix. On the other hand, the occurrence of small voids also can be clearly seen on the fractured surface of 3 phr and 5 phr copper (I) oxide added LDPE composites when irradiated up to 300 kGy. The appearance of voids is mainly attributed to the gas released during the chain scissioning process which these voids also could further reduce the matrix continuities of LDPE samples.

X-Ray Diffraction Analysis

The XRD patterns of pristine LDPE and LDPE composites added with various loading levels of copper (I) oxide when irradiated at various irradiation dosages are shown in Figure 6. By referring to Figure 6(a), two significant diffraction peaks which are angles deflection (110) and (020) were observed to appear at $2\theta = 21.70^o$ and 23.93^o, respectively in the XRD pattern of un-irradiated pristine LDPE. On the other hand, three diffraction peaks of (110), (200) and (020) can be observed to occur on the XRD curves of LDPE composites added with copper (I) oxide at $2\theta \approx 21.5^o$, 23.7^o and 36.4^o, respectively [34]. The increasing of copper (I) oxide loading level could significantly induce the peak intensity for deflection peak of (020) on XRD curves of LDPE composites as shown in Figure 6(a). The increment in peak intensity represents the convergence of random polymer chains arrangement structures in LDPE matrix towards more ordered arrangement after the addition of copper (I) oxide (Kumar et al., 2012) [13]. Besides, the deflection peak (020) on XRD curves also observed to be sharper (narrower in peak width) when the loading level of copper (I) oxide added in LDPE matrix was gradually increased. The sharpening of deflection peak (020) is associated with the increment in the average crystallite size [13]. However, the peak intensity of deflection peaks (110) and (200) was found to be significantly reduce with increasing of copper (I) oxide loading level as illustrated in Figure 6(a). This indicates the addition of copper (I) oxide into LDPE matrix could disturb and slightly rupture the highly ordered chains arrangement inside polymer matrix into random chains arrangement. The broadening of deflection peaks (110) and (200) also can be observed in Figure 6(a). This also indicates that the increasing of copper (I) oxide loading level could cause the reduction in average crystallite size.

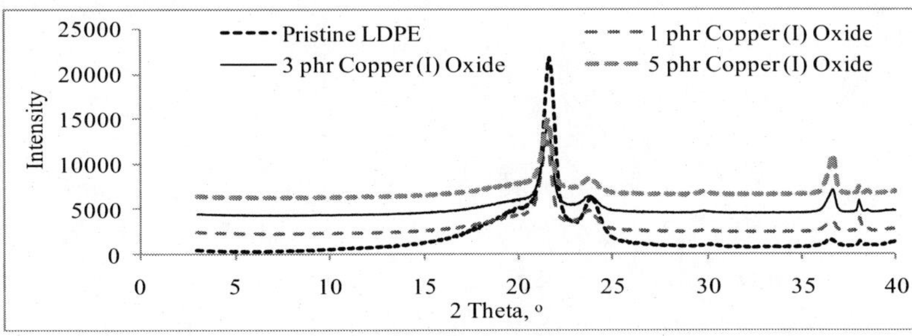

(a) Non-irradiated

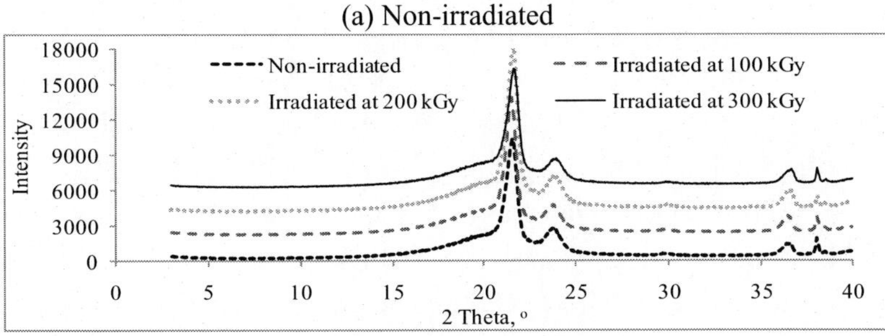

(b)1 phr Copper (I) oxide

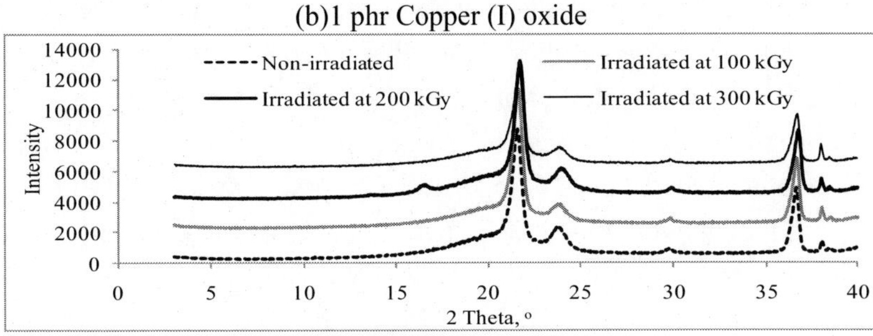

(c) 5 phr Copper (I) oxide

Figure 6. XRD pattern curves of (a) un-irradiated LDPE composites added with increasing of copper (I) oxide loading levels, (b) 1 phr copper (I) oxide added LDPE composites when subjected to various irradiation dosages and (c) 5 phr copper (I) oxide added LDPE composites when subjected to various irradiation dosages.

By referring to Figures 6(b) and 6(c), the peak intensity of deflection peak (020) on XRD curves of all non-irradiated and irradiated 5 phr copper (I) oxide (high loading level) added LDPE composites was observed to be significantly higher than 1 phr copper (I) oxide (low loading level) added LDPE composites. The deflection peak (020) of all 5 phr copper (I) oxide added LDPE composites also observed to be sharper than all 1 phr copper (I) oxide added LDPE composites. However, the peak intensity of deflection peak (110) on XRD curves of all 5 phr copper (I) oxide added LDPE composites is found to be lower than all 1 phr copper (I) oxide added LDPE composites.

Figure 7 illustrates the effect of increasing the copper (I) oxide loading level and the electron beam irradiation dosages on the crystallinity of pristine LDPE and all copper (I) oxide added LDPE composites. By referring to Figure 7(a), the crystallinity of un-irradiated LDPE was found to be gradually increased with increasing of copper (I) oxide loading level. This indicates the addition of copper (I) oxide into LDPE matrix could contribute in converging the polymer chains structure arrangement in LDPE matrix into more ordered polymer chains arrangement. This might be due to the addition of higher loading level of copper (I) oxide (3 phr and 5 phr) could induce the degree of regularity in the polymer chains arrangement and ordering structure of LDPE matrix [35]. Under electron beam irradiation dosage of 100 kGy, identical trend also can be noticed on the crystallinity results of LDPE composites added with increasing loading level of copper (I) oxide as shown in Figure 7(a). By referring to Figures 7(a) and 7(b), the application of electron beam irradiation dosage up to 100 kGy has significantly increased the crystallinity of pristine LDPE and all copper (I) oxide added LDPE composites. This also indicates the formation of crosslinking networks in LDPE matrix by electron beam irradiation could induce the degree of regularity in polymer chains and copper (I) oxide arrangement structures inside LDPE matrix. For LDPE composites added with 1 phr and 3 phr copper (I) oxide, the crystallinity was found to be increase with increasing of electron beam irradiation dosage up to 200 kGy as illustrated in Figure 7(b). This also indicates that the formation of crosslinking networks under irradiation dosage of 200 kGy could further promote the structural rearrangement of polymer chains and copper (I) oxide particles into higher ordered chains arrangement [13]. However, the crystallinity of LDPE composites added with increasing of copper (I) oxide loading level was observed to significantly reduce when subjected to higher electron beam irradiation dosage up to 300 kGy. At higher irradiation dosage up to 300 kGy, the chains scissioning process is more pre-dominate to occur than the crosslinking process. The chains scissioning process could attack and breakdown the backbone of polymer chains into smaller chains. The breakdown of polymer chains into smaller chains could rupture the highly ordered polymer chains arrangement inside LDPE matrix and reduce the crystallinity of LDPE composites [35].

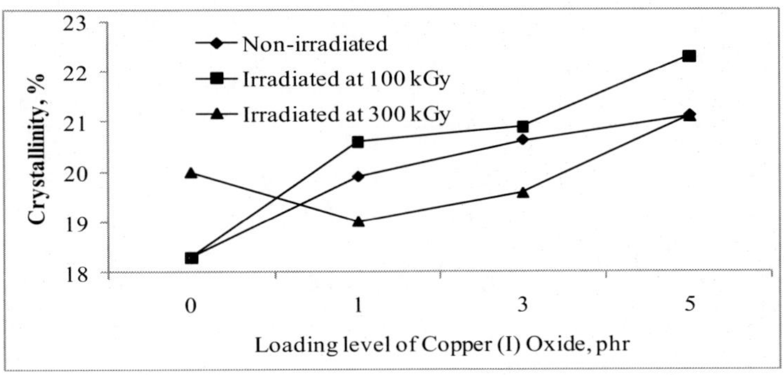

(a)

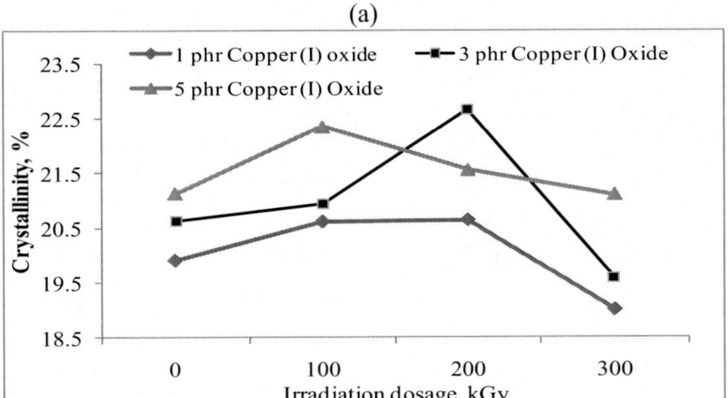

(b)

Figure 7. Effect of (a) increasing copper (I) oxide loading level and (b) increasing electron beam irradiation dosages on the crystallinity of irradiated-copper (I) oxide added LDPE composites.

Table 2. The angle, crystallite size, d-spacing and interchain separation of deflection peak (110) for all copper (I) oxide added LDPE composites under various irradiation dosages

Deflection peak (110) on XRD curves in Figure 6					
Loading level of copper (I) oxide, phr	Irradiation dosage, kGy	Angle (2θ), °	Crystallite size (*L*), Å	d-spacing, Å	Interchain Separation (*R*), Å
0	0	21.7000	136.9	4.091	5.113
	100	21.3900	153.1	4.149	5.186
	200	21.3801	126.4	4.151	5.189
	300	21.5768	120.3	4.114	5.142
1	0	21.4787	117.7	4.132	5.165
	100	21.4420	124.4	4.139	5.173
	200	21.5356	125.1	4.121	5.152
	300	21.5622	128.8	4.116	5.145
3	0	21.5547	129.7	4.118	5.147
	100	21.5551	130.3	4.124	5.155
	200	21.5045	142.6	4.127	5.159
	300	21.5326	114.1	4.122	5.153
5	0	21.5386	133.2	4.121	5.151
	100	21.5817	142.5	4.113	5.141
	200	21.5836	139.7	4.112	5.140
	300	21.6134	130.7	4.107	5.133

The deflection peaks with highest intensity for all LDPE composites added with increasing loading level of copper (I) oxide are observed to be deflection peak (110), (200) and (020), respectively as shown in Figure 6. According to Tables 2-4, the crystallite size, d-spacing and interchain separation for all copper (I) oxide LDPE composites under various irradiation dosages were calculated in corresponding to these deflection peaks. The addition of 1 phr copper (I) oxide particles into LDPE matrix was observed to highly decrease the crystallite size of deflection peaks (110) and (200). The addition of low copper (I) oxide loading level (1 phr) into LDPE matrix could significantly reduce the original large size crystallite in amorphous regions of LDPE matrix by disturbing and rupturing the highly ordered chains arrangement in crystallite in LDPE matrix. The rupture of the existing crystallite structures could significantly reduce the crystallite size of deflection peaks (110) and (200). However, the crystallite size of deflection peaks (110), (200) and (020) was found to be gradually increased as the loading level of copper (I) oxide increased from 1 phr to 5 phr as depicted in Tables 2-4. This is attributable to the increasing of copper (I) oxide amounts (≥3 phr) in LDPE matrix could promote the formation of highly ordered arrangement structures by forming new crystallite structures [13, 36]. The newly formed crystallite structure would interact with the original large size crystallite in LDPE matrix and thus increase the crystallite size of deflection peaks of (110) and (200). By referring to Tables 2-3, the deflection peaks (110) and (200) of pristine LDPE was found to marginally increased when irradiated up to 100 kGy. The increment in crystallite size is attributed to the formation of new crystallites in amorphous regions after subjected to electron beam irradiation. The interaction between the newly formed crystallites and original large size crystallite could significantly increase the crystallite size of crystallites (110) and (200). However, the crystallite size for crystallites (110) and (200) was gradually reduced with further increasing the irradiation dosage from 100 kGy to 300 kGy as shown in Tables 2-3. The reduction in crystallite size indicating the application of higher irradiation dosages (≥200 kGy) could rupture the crystallites structures by disturbing the highly ordered chains arrangement into more random chains arrangement. In other words, the reduction in crystallite size might be attributed to the occurrence of chain scissioning process at higher irradiation dosages which could breakdown the large macromolecules chains into smaller chains [36]. The breakdown of macromolecular chains could disturb the highly ordered chains arrangement into less ordered. At lower loading level of copper (I) oxide (1 phr and 3 phr), the application of electron beam irradiation dosage up to 200 kGy has gradually increased the crystallite size of crystallites (110), (200) and (020). This might be attributed to the addition of low copper (I) oxide loading level (≤3 phr) could further induce the formation of crosslinking networks in LDPE matrix under electron beam irradiation [8]. The formation of crosslinking networks in LDPE matrix could promote the formation of some crystallites in the amorphous regions of LDPE matrix by rearranging to more ordered chains arrangement [13]. These newly formed crystallites would interact with the existing (original) crystallites and cause the crystallite size of crystallites (110) and (200) to significantly increase. However, the crystallite size of all crystallites was observed to be decrease when further irradiated up to 300 kGy. This is due to the chains scissioning process is more likely to occur when subjected to higher irradiation dosage (≥300 kGy). During the chains scissioning process, generated free radicals tended to attack the backbone chains of LDPE macromolecules and breakdown the macromolecules chains of LDPE into smaller chains. Thus, this process could significantly disturb the ordered chains arrangement in LDPE matrix and reduce the crystallite size of these crystallites.

Table 3. The angle, crystallite size, d-spacing and interchain separation of deflection peak (200) for all copper (I) oxide added LDPE composites under various irradiation dosages

Deflection peak (200) XRD curves in Figure 6					
Loading level of copper (I) oxide, phr	Irradiation dosage, kGy	Angle (2θ), °	Crystallite size (*L*), Å	d-spacing, Å	Interchain Separation (*R*), Å
0	0	23.9301	103.1	3.714	4.643
	100	23.6500	105.4	3.758	4.697
	200	23.6202	92.9	3.762	4.703
	300	23.8407	88.5	3.728	4.660
1	0	23.6956	82.3	3.750	4.688
	100	23.6751	86.0	3.754	4.692
	200	23.7928	86.6	3.735	4.669
	300	23.7918	78.7	3.736	4.669
3	0	23.7748	85.2	3.738	4.673
	100	23.8202	89.0	3.731	4.664
	200	23.7412	91.5	3.743	4.679
	300	23.7298	78.3	3.736	4.669
5	0	23.7675	89.6	3.739	4.674
	100	23.7571	91.2	3.741	4.676
	200	23.7781	84.2	3.738	4.672
	300	23.7754	80.8	3.738	4.673

Table 4. The angle, crystallite size, d-spacing and interchain separation of deflection peak (020) for all copper (I) oxide added LDPE composites under various irradiation dosages

Deflection peak (020) on XRD curves in Figure 6					
Loading level of copper (I) oxide, phr	Irradiation dosage, kGy	2 Theta (2θ), °	Crystallite size (*L*), Å	d-spacing, Å	Interchain Separation (*R*), Å
0	0	-*	-*	-*	-*
	100	-*	-*	-*	-*
	200	-*	-*	-*	-*
	300	-*	-*	-*	-*
1	0	36.3954	140.2	2.466	3.082
	100	36.3610	148.4	2.468	3.085
	200	36.4116	141.1	2.465	3.081
	300	36.4419	133.2	2.463	3.078
3	0	36.4910	170.0	2.459	3.074
	100	36.4902	170.0	2.459	3.074
	200	36.4421	183.3	2.463	3.078
	300	36.5106	160.0	2.458	3.073
5	0	36.5066	188.2	2.458	3.073
	100	36.5280	188.4	2.457	3.071
	200	36.5384	190.6	2.456	3.070
	300	36.5652	178.7	2.455	3.068

* = no peak is detected on XRD pattern curves.

By referring to Tables 2-3, the addition of copper (I) oxide particles in LDPE matrix was observed to slightly increase the d-spacing and interchain separation of crystallites (110) and (200). The increment in d-spacing and interchain separation of these crystallites indicates that the polymer chains in these crystallites become less compact after the addition of copper (I) oxide particles. In other words, the copper (I) oxide particles could disperse into the polymer chains in crystallites and the compactness of the polymer chains (crystal planes) has been reduced [35]. For pristine LDPE, the application of electron beam irradiation up to 200 kGy was found to marginally increase the d-spacing and interchain separation of crystallite (110) and (200). This also indicates that the formation of crosslinking networks induced by electron beam irradiation could significantly reduce the compactness of polymer chains in LDPE matrix by increasing the macromolecular size of polymer chains. However, further increasing in irradiation dosage from 200 kGy to 300 kGy has significantly reduced the d-spacing and interchain separation of these crystallites. This could be attributable to the higher irradiation dosage up to 300 kGy could breakdown the large and long chains in crystallites into smaller and shorter chains through the chain scissioning process. The breakdown of large and long polymer chains into smaller and shorter chains has significantly reduced the density of macromolecules chains in crystallites. This could cause the shorter polymer chains in crystallites to align in more close to each other and the compactness of these crystallites also increased [35]. Thus, the d-spacing and interchain separation has been significantly reduced. Similar observation also can be found in the samples of 1 phr copper (I) oxide added LDPE composites (lower loading level) as shown in Tables 2-3. The application of electron beam irradiation dosages up to 100 kGy could slightly increase the d-spacing and interchains separation of crystallites (110) and (200). However, the d-spacing and interchain separation of these crystallites were significantly reduced when further irradiated up to 200 kGy and 300 kGy. This is due to the occurrence of chains scissioning process at higher irradiation dosages which could reduce the size of polymer chains in crystallites and thus increase the compactness of polymer chains aligned in these crystallites. However, the application of electron beam irradiation was observed to pose insignificantly change in d-spacing and interchain separation of 1 phr copper (I) oxide added LDPE composites. This is due to the occurrence of crystallite (020) in LDPE matrix is mainly attributed to the addition of copper (I) oxide particles, which the polymer chains in LDPE matrix unable to intercalate into crystal planes of copper (I) oxide particles [20]. Thus, the d-spacing and interchain separation of the crystal planes for crystallite (020) did not provide any significant change. Besides, the effect of electron beam irradiation on the d-spacing and interchain separation of crystallites (110), (200) and (020) for higher loading level of copper (I) oxide added LDPE composites (3 phr and 5 phr) was observed to be less significant.

CONCLUSION

This study has investigated the effect of adding copper (I) oxide on physico-mechanical characteristics of electron beam irradiated LDPE composites. The addition of lower loading level of copper (I) oxide (≤1 phr) into LDPE matrix have significantly decreased the gel content of irradiated LDPE composites as compared to pristine LDPE. This is due to the addition of copper (I) oxide in low amounts could block and hinder the formation of crosslinking networks in LDPE matrix. At higher loading level of copper (I) oxide (≥3 phr), the gel content of LDPE composites has been marginally increased. This is due to the higher amounts of copper (I) oxide particles in LDPE matrix could induce the formation of polymeric free radicals in LDPE matrix by absorbing the accelerated electron from electron beam accelerator. Thus, the increasing of copper (I) oxide loading level could further increase the degree of crosslinking networks formed in LDPE matrix. On the other hand, the incorporation of copper (I) oxide into LDPE matrix has severely decreased the tensile strength and elongation at break of LDPE matrix. The addition of copper (I) oxide particles with irregularity shapes into LDPE matrix could disrupt the structure arrangement of LDPE matrix by further reducing the intermolecular bonding between the LDPE macromolecular chains. However, the increasing of copper (I) oxide loading level up to 3 phr has significantly increased the tensile strength of LDPE composites by providing the reinforcement effect to polymer matrix. The higher loading level of copper (I) oxide could reduce the irregularity arrangement of copper (I) oxide in LDPE matrix by homogenously distributing the copper (I) oxide particles in LDPE matrix. The homogenous dispersion of copper (I) oxide particles in LDPE matrix could enhance the interfacial adhesion effect between copper (I) oxide particles and LDPE matrix, thus the applied stress on polymer matrix can be effectively transferred throughout the polymer matrix. However, further increment in copper (I) oxide loading level has reduced the tensile strength of LDPE composites. The copper (I) oxide particles tended to agglomerate together into larger and irregular agglomerated particles at higher loading level as observed in SEM analysis could reduce the interfacial adhesion effect between copper (I) oxide and LDPE matrix, thus reducing the tensile strength and elongation at break. The application of electron beam irradiation dosages up to 150 kGy has significantly improved the tensile strength of all copper (I) oxide added LDPE composites by introducing the formation of crosslinking networks in LDPE matrix. The formation of crosslinking networks could improve the structural arrangement of copper (I) oxide in LDPE matrix by enhancing the interfacial adhesion effect between coper (I) oxide particles and LDPE matrix. The increasing of copper (I) oxide loading level has gradually increased the crystallinity of LDPE matrix by converging the random chains arrangement in amorphous regions into more ordered chains arrangement. The application of irradiation dosage up to 100 kGy and 200 kGy also significantly increased the crystallinity of copper (I) oxide added LDPE composites by the formation of crosslinking networks. The formation of crosslinking networks in LDPE matrix could rearrange the random chains structures into highly ordered chains arrangement. On the other hand, the crystallite size of all crystallites for pristine LDPE and all copper (I) oxide LDPE composites was significantly increased when irradiated up to 100 kGy and 200 kGy. This is due to the electron beam irradiation could induce the formation of new crystallites in amorphous regions where these newly formed crystallite structures would interact well with the original crystallite structures. Besides, the application of electron beam irradiation dosages has marginally increased the d-spacing and interchain separation of all crystallites which indicating the formation of crosslinking networks could reduce the compactness of polymer chains in these crystallites.

ACKNOWLEDGMENT

The authors are very appreciating with the kindliness of Malaysian Nuclear Agency, Bangi, Selangor for allowing usage of their equipments on performing this research.

REFERENCES

[1] Briscoe BJ, Rodosian AK, Tabor D. The friction and wear of high density polyethylene: the action of lead oxide and copper oxide fillers. *Wear* 1974;27: 19-34.

[2] Bahadur S, Schwartz C. The effect of nanoparticle fillers on transfer film formation and the tribological behaviour of polymer. In: Friedrich K, Schlarb A K, editor. *Tribology of Polymeric Nanocomposites Friction and Wear of Bulk Materials and Coatings* 2nd Ed. United Kingdom: Butterworth-Heinemann, 2013. p. 23-48.

[3] Bahadur S., Tabor D. Role of fillers in the friction and wear behaviour of high density polyethylene. In: Lee, L-H, editor. *Polymer Wear and Its Control*, ACS Symp. Ser. 54th Ed. Washington: American Chemical Society, 1985. p. 253-268.

[4] Bahadur S., Tabor D. The wear of filled polytetrafluoroethylene. *Wear* 1984;98: 1-13.

[5] Gong D, Xue Q, Wang H. Study of the wear of filled polytetrafluoroethylene. *Wear* 1989;134: 283-295.

[6] Palabiyi M, Bahadur S. Tribological studies of polyamide 6 and high-density polyethylene blends filled with PTFE and copper oxide and reinforced with short glass fibers. *Wear* 2002;253: 369-376.

[7] Bahadur S and Gong D. The role of copper compounds as fillers in transfer film formation and wear of nylon. *Wear* 1992;154: 207-223.

[8] Osawa Z. Role of metals and metal-deactivators in polymer degradation. *Polym Degrad Stab* 1988;20: 203-236.

[9] Abliz D, Duan Y, Zhao X, Li D. Low-energy electron beam cured tape placement for out-of-autoclave fabrication of advanced polymer composites. *Compos Part A: Appl Sci Manuf* 2014; 65: 73-82.

[10] Xu K, Li K, Zhong T, Guan L, Xie C, Li S. Effects of chitosan as biopolymer coupling agent on the thermal and rheological properties of polyvinyl chloride/wood flour composites. *Compos B: Eng* 2014;58: 392-399.

[11] Dalai S, Wenxiu C. Radiation effects on LDPE/EVA blends. *Appl Polym Sci* 2002;86: 1296-1302.

[12] Murray KA, Kennedy JE, McEvoy B, Vrain O, Ryan D, Cowman R, Higginbotham CL. The effects of high energy electron beam irradiation in air on accelerated aging and on the structure property relationships of low density polyethylene. *Nucl Instrum Methods Phys Res B* 2013;297: 64-74.

[13] Raghavan J. Evolution of cure, mechanical properties, and residual stress during electron beam curing of a polymer composite. *Compos Part A: Appl Sci Manuf* 2009;40: 300-308.

[14] Dadbin S, Frounchi, Saeid MH, Gangi F. Molecular Structure and Physical Properties of E-Beam Crosslinked Low-Density Polyethylene for Wire and Cable Insulation Applications. *J Appl Polym Sci* 2002;86: 1959-1969.

[15] Rytlewski P, Malinowski R, Moraczewski K, Zenkiewicz M. Influence of some cross-linking agents on thermal and mechanical properties of electron beam irradiated polylactide. *Radiat Phys Chem* 2010;79: 1052–1057.

[16] Khonakdar HA, Jafari SH, Wagenknecht U, Jehnichen D. Effect of electron-irradiation on cross-link density and crystalline structure of low- and high-density polyethylene. *Radiat Phys Chem* 2006;75: 78–86.

[17] Modesti M, Lorenzetti A. Halogen-free flame retardants for polymeric foams. *Polym Degrad Stab* 2002;78: 167-173.

[18] Bhattacharya A. Radiation and Industrial polymers. *Prog Polym Sci* 2000;25: 371-401.

[19] Bee ST, Hassan A, Ratnam CT, Tee TT, Sin LT. Effects of montmorillonite on the electron beam irradiated alumina trihydrate added polyethylene and ethylene vinyl acetate nanocomposites. *Polym Compos* 2012;33: 1883-1892.

[20] Bee ST, Hassan A, Ratnam CT, Tee TT, Sin LT. Investigation of nano-size montmorillonite on electron beam irradiated flame retardant polyethylene and ethylene vinyl acetate blends. Nucl. Instrum. *Methods Phys Res* B 2013;299: 42–50.

[21] Bee ST, Hassan A, Ratnam CT, Tee TT, Sin LT, Hui D. Dispersion and roles of montmorillonite on structural, flammability, thermal and mechanical behaviours of electron beam irradiated flame retarded nanocomposite. *Comp Part B: Eng* 2014; 61: 41-48.

[22] Basfar AA, Ali ZI. Physico-chemical properties of low density polyethylene and ethylene vinyl acetate composites cross-linked by ionizing radiation. *Radiat Phys Chem* 2011; 80: 257-263.

[23] Liu Z, Jin J, Chen S, Zhang J. Effect of crystal form and particle size of titanium dioxide on the photodegradation behaviour of ethylene-vinyl acetate copolymer/low density polyethylene composite. *Polym Degrad Stab* 2011; 96: 43-50.

[24] Chattopadhyay S, Chaki TK, Bhowmick AK. Electron beam modification of the thermoplastic elastomeric blends made from polyolefins. *J Mater Sci* 2001; 36: 4323-4330.

[25] Cárdenas MA, García-López D, Gobernado-Mitre I, Merino JC, Pastor JM, Martínez JD, Barbeta J and Calveras D. Mechanical and fire retardant properties of EVA/clay/ATH nanocomposites-Effect of particle size and surface treatment of ATH filler. *Polym Degrad Stab* 2008; 93: 2032-2037.

[26] Hong SG, Yeh CS. The effects of copper oxides on the thermal degradation of bismaleimide triazine prepreg. *Polym Degrad Stab* 2004; 83: 529-537.

[27] Basfar AA. Flammability of radiation cross-linked low density polyethylene as an insulating material for wire and cable. *Radiat Phys Chem* 2002; 63: 505-508.

[28] Sharif J, Mohd Dahlan KZ, Wan Yunus WMZ. Radiation crosslinked natural rubber/ethylene vinyl acetate/clay nanocomposites-Effect of organoclay concentration. *J Nuc Relat Techn* 2008; 5(1): 65-80.

[29] Murray KA, Kennedy JE, McEvoy B, Vrain O, Ryan D, Higginbotham CL. The effects of high energy electron beam irradiation on the thermal and structural properties of low density polyethylene. *Radiat Phys Chem* 2012; 81(8): 962-966.

[30] Haurie L, Fernández AI, Velasco JI, Chimenos JM, Cuesta JML, Espiell F. Thermal stability and flame retardancy of LDPE/EVA blends filled with synthetic hydromagnesite/aluminium hydroxide/ montmorillonite and magnesium hydroxide/aluminium hydroxide/ montmorillonite mixtures. *Polym Degrad Stab* 2007; 92: 1082-1087.

[31] Munusamy Y, Ismail H, Mariatti M, Ratnam CT. Effect of Electron Beam Irradiation on the Properties of Ethylene-(vinyl acetate) Copolymer/Natural Rubber/Organoclay Nanocomposites. *J Vinyl Addit Techn* 2009; 15: 39-46.

[32] Sharif J, Abdul Aziz SHS, Hashim K. Radiation effects on LDPE/EVA blends. *Radiat Phys Chem* 2000; 58: 191-195.

[33] Sharif J, Mohd Dahlan KZ and Wan Yunus WMZ. Electron beam crosslinking of poly(ethylene-co-vinyl acetate)/clay nanocomposites. *Radiat Phys Chem* 2007; 76: 1698-1702.

[34] Baker AME, Windle AH. Evidence for a partially ordered component in polyethylene from wide-angle X-ray diffraction. *Polymer* 2001; 42: 667-680.

[35] Bhadra S, Khastgir D. Determination of crystal structure of polyaniline and substituted polyanilines through powder X-ray diffraction analysis. *Polymer Testing* 2008; 27: 851-857.

[36] Bee ST, Ratnam CT, Sin LT, Tee TT, Wong WK, Lee JX, Rahmat AR. Effects of electron beam irradiation on the structural properties of polylactic acid/polyethylene blends. *Nuc Inst Meth Phys Res B* 2014; 334: 18-27.

In: Polymer-Matrix Composites ISBN: 978-1-53613-398-1
Editors: Elias Rice and Beth Sparks

Chapter 3

THERMAL AGEING STUDIES OF STARCH-BASED NANOCOMPOSITES REINFORCED WITH DIFFERENT WEIGHT RATIO ORGANIC-TO-INORGANIC NANOFILLERS

Hon-Meng Ng*[1]*, Lee Tin Sin*[1,*]*, Tiam-Ting Tee*[1]*,
***Soo-Tueen Bee*[1,†] *and A. R. Rahmat*[2]**
[1]Department of Chemical Engineering,
Lee Kong Chian Faculty of Engineering and Science,
Universiti Tunku Abdul Rahman,
Bandar Sungai Long, Jalan Sungai Long, Kajang, Malaysia
[2]Department of Polymer Engineering,
Faculty of Chemical Engineering, Universiti Teknologi Malaysia,
UTM Skudai, Johor

* Corresponding Author Email: direct.tinsin@gmail.com.
† Corresponding Author Email: direct.beest@gmail.com.

ABSTRACT

The inorganic Montmorillonite (MMT) and organic cellulose nanofibers (CNF) obtained through acid hydrolysis were simultaneously added to plasticized starch with a defined CNF-to-MMT ratio and processed into nanocomposites by melt extrusion. In order to investigate the possibility of organic nanofillers an alternative to inorganic nanofillers in sustaining the composite properties for food container application under thermal exposures, the continuously isothermal aerobic ageing on samples was carried out by oven-heating at 60°C. Measurements of tensile properties, impact strength, water absorption and cross-sectional morphology were conducted on TPS and nanocomposites as a function of aging time (0-, 6-, 12-, 18-, 24-, 48- and 72-hours). The changes of internal structure or performance of studied materials varied with the different ageing stages. The tensile results elucidated the increase of material compactness by formation of crosslinking during initial ageing to decrease the chain mobility. As compared to MMT, CNF is more suitable to serve as a post cure agent since MMT has inhibited the increment of TS. The strength/modulus are found to be decreased during middle ageing stage which related to the degradation induced by intensive thermal exposure, but the ensuing molecular rearrangement of scissioned remainders tend to recover properties. Changes in morphology of fractured samples after thermal aging revealed the modification of TPS morphology fracture surface by crosslinking network. The obtained degree of crosslinking density could be reflected by % water absorption. The impact loading response manifested the high brittleness of thermal-aged-samples based on unfavourably deteriorated impact resistance.

Keywords: polymer-matrix composites (PMCs), mechanical properties, thermal analysis, nanofillers

1. INTRODUCTION

Consciously, alarming plastic wastes of conventional food container highly probable to become one of the most prominent issues in the near future years, as realized from the increasingly discarded amount lately upon the decay in the usefulness [1]. With good environmentally biocompatibility, thermoplastic starch (TPS) is of interest to be possible biopolymer feedstocks for food containers, while this movement has been inspired by already use of starch to derive composite foam for single-use food service items [2, 3]. Indeed, the biocompatible manner of starch is originated from inherently biodegradable backbones (hydrolysable acetal/glycosidic link) along amylose and amylopectin, which leading biopolymers to be availably degraded by the enzymatic action of living organisms [4, 5]. Under heating and shearing effects, plasticized starch can be processed into a moldable thermoplastic through starch gelatinization, known as thermoplastic starch (TPS) [6]. During the thermoplastic processing, it has been commonly used stearic acid as the additive for polymeric starchy materials since fatty acid chain with little internal flexibility could play an indispensably lubricating role in facilitating the smooth processing [6, 7]. Critically, the properties of pure starch-based thermoplastics are still unsatisfactory for such potential movement based on their poor long-term stability, high water sensitivity (swollen by water) and weak mechanical property [8, 9]. These facts stimulate interests in synthesizing composite by the addition of fillers or reinforcements, to overcome these drawbacks TPS [8].

Fabrication of composites based on organic or inorganic reinforcement is another possible solution to improve the performance of starch films. Application of fillers such as clays and fibers have been reported as promising candidates for starch reinforcement while much attentions has been focused on the manipulation of fillers into nanosize with dimension of nanometer range, i.e., 1–100 nm [9, 10]. The montmorillonite (MMT) has been regarded as representative nanoclay in reinforcing TPS based on montmorillonite crystal lattice is layered silicates consisting of platelets with high aspect ratio, enabling to induce intercalation/exfoliation reinforcing system [11]. Physically, MMT is so-called nanofiller mainly attributed to its stack of two-dimensional (2D) nanosheets which is 1 nm in thickness for each alumino-silicate layers and 200–300 nm in the lateral dimension [12, 13]. Within gallery space between the aluminosilicate layers, the negatively charged silicate layer is surface-substituted with alkali cations (e.g., Cloisite Na^+) [14]. Another nanofiller used to reinforce TPS is plant-derived cellulose nanofiber (CNF) while cellulose is composed of β-1,4 linked glucopyranose units, with polymer chains associated by hydrogen bonds forming bundles of fibrils [15]. CNF are generally divided into nanofibrillated cellulose and cellulose nanocrystals (CNC) depending on extraction techniques from plant sources. When CNF is mixed with starch, CNF was able to improve the functional properties of starch-based films. This fact has been attributed to the high chemical similarity of starch and CNF, providing good compatibility and interaction between them [16]. Targeted on reinforcing TPS, CNF and MMT comparably possess similar reinforcing potential based on their nano-scale effect and hydrophilic character, as evidenced by their similar improvement of tensile strength of TPS to approximately 100% at certain loading level (~10 wt.%), according to Hietala et al., (2013) [17] and Ning et al., (2009) [18].

After being reinforced by nanofiller, it is considerably desired to develop reinforced-starch as disposable household food container without leaching the harmful substance into environment like conventional plastic container. Since sustainability of product's properties is regarded as priority of service life, an issue is possibly raised concerning to the stability of nanocomposites' performance under long-term exposure to high temperature. It should be realizable that the ambient high temperature could be come from loading hot cooked food, thus it is of interest to comprehend the thermal aging behavior of synthesized food container when considering implementation. In general, the effect of thermal aging on polymer is a combination of consolidation (crosslinking) and degradation (chain-scission) mechanisms or oxidative degradation [19]. The thermal ageing of polymer has been investigated as an efficient route to improve the properties lies in the heat stresses from thermal energy is contributable to liberate free radicals for initiating crosslinking between the generating radicals R• (molecules with an unpaired electron). Cross-linking usually yields the increase in modulus and the decrease in extensibility of the material. In many cases, thermal degradation and oxidation are two important ageing mechanisms once thermal-oxidative ageing is initiated by high enough heat energy [20]. The thermooxidative ageing would be developed on the heated solid phase (polymer) which tends to react with ambient gas (oxygen) under high temperature (mostly below the melting temperature) and appropriate oxygen partial pressure mainly investigating [21]. The autooxidation starts by oxygen in its diradical triplet-shape |Ó-Ó| ≡ |O=O| immediately reacts with generated alkyl radicals (R•) into the peroxy radical ROO•. The further reaction of the peroxy radical with RH formed into hydroperoxide ROOH is rate determining [21, 22]. Diffusion and solubility of O_2 are material depending and influence the oxidation reaction significantly [21, 22].

In this paper, one can highlight that the cellulose nanofiber was prepared from kenaf fiber without undergoing typical dialysis after sulfuric acid treatment, instead of the acid sulfate groups are balanced by NaOH solution. This is based on a research paper stated that the neutralization after hydrolysis gives rise to higher increase in the thermal stability of CNCs than dialysis [23]. With attractive biocompatibility and biodegradability, the derived CNF much more desired to maintain one of the most important advantages of starch (i.e., biodegradability) than inorganic MMT [24, 25]. From the view point of sustainable production, CNF has considerably greater priority to serve as relevant reinforcing agent as realized from the possibility of mineral MMT to be exhausted in one day of future based on non-renewable behaviour of clay mineral on earth. Thus, this research has attempted to replace nanoclay with renewable nanofiber in order to investigate CNF as an alternative to improve the properties of TPS as MMT [26]. In addition, the successfully entire replacement of MMT with CNF desired to save the limited clay resources and the cost of the end-products considering the relatively higher cost of MMT. As a consequence, the present research work is also centered on the investigation of CNF as possible alternative to MMT in sustaining the properties of bionanocomposite under the condition of thermal-ageing performed by laboratory accelerated tests. In order to deal with the comparison of reinforcing potential between CNF and MMT, the MMT incorporated in starch is progressively substituted by corresponding amount CNF, resulting in various filler ratio (CNF:MMT = 0:10, 2:8, 4:6, 6:4, 8:2, 10:0) in starch matrix. The properties of of aged samples were determined in the same manner as the non-aged samples by tensile test, impact test, scanning electron microscopy (SEM) and water absorption test.

2. Experimental

2.1. Materials

The kenaf bast fiber for potential use as reinforcement fibers in the composite was supplied by the local sources of J&2 Agrofarm, Kulai, Johor, Malaysia. The chemicals (NaOH pellets, glacial acetic acid (99–100%), sodium chlorite ($NaClO_2$, technical grade, 80%, Sigma-Aldrich) and sulfuric acid (95.0–98.0 wt%,) which used for cellulose nanofiber preparation were of reagent grade and purchased from Genscience Private Ltd., Malaysia and Fisher Scientific Private Ltd., Malaysia respectively. Inorganic nanofiller, Na-MMT (Nanoclay brand-Nanolin DK) with nanosize of 1-100nm was purchased from Sino-Holding Chemicals Co., Ltd. Tapioca starch, Glycerol purified (Glycerine) and stearic acid (granular form) used for processing of based polymer was purchased from Mydin supermarket and Symphony Private Ltd., Malaysia.

2.2. Extraction of Cellulose Nanofibers

Cellulose nanofibers were extracted from kenaf bast fiber by a series of chemical treatment, followed by some post treatments. Initially, the raw kenaf fiber was ground with a universal cutting mill (pulverisette, FRITSCH) to pass through a 60-mesh (250 μm opening) within sieve vibrator. In the chemical pre-treatment stage, the cut fibers (Figure 1a) were firstly treated with a sodium hydroxide (NaOH) aqueous solution of 4% (w/w) at 80 ± 5ºC under mechanic stirring for 2 h. The alkali treatment was performed four times to remove most non-cellulosic materials (hemicellulose, lignin) and some impurities (waxy, for instance), and finally dried at 40°C for 24 h in an air-circulating oven after being washed.

The alkali treated pulp (Figure 1b) was bleached with a solution made up of equal parts (v:v) of acetate buffer (27 g NaOH and 75 mL glacial acetic acid, diluted to 1 L of distilled water) and aqueous sodium chlorite (1.7 wt.% $NaClO_2$ in water) at 75 ± 5ºC under mechanic stirring for 2 h. This bleaching treatment was repeated three times to remove lignin and remaining hemicellulose until the product became white and subsequently washed repeatedly in distilled water until the neutral pH was reached before being dried at 40°C for 24 h in an air-circulating oven. The cellulose nanocrystals was obtained by sulfuric acid hydrolysis (65 wt% H_2SO_4 solution, 45°C, 40 min) of bleached fibers (Figure 1c) under vigorous stirring resulting in a stable colloidal suspension. The acid-hydrolyzed precipitate in form of viscous gel (Figure 1d) were separated by centrifugation (7000 rpm, 5 min, 4°C) and washed by re-dispersing in water, this cycle were performed 3 times. The remaining acid-sulphate groups which has been identified to deteriorate thermal-stability of cellulosic substances would be neutralized by diluted NaOH solution (1 wt %) followed by introducing in ultrasonic bath for 1 hour to obtain a final aqueous suspension of cellulose nanofibers which does not sediment or flocculate. In order to obtain dried CNF (Figure 1e), the aqueous CNF suspension was frozen by refrigerator at -24°C overnight before being subjected to freeze drying (0.14 mBar, -40°C) for 48 hours. Finally, the nanosize of extracted cellulose nanofibers was examined by subjecting the acid-hydrolyzed product to Atomic Force Microscope (NanoScope IIIa AFM, NSC14 series, MikroMasch) in tapping mode using Ultrasharp Si tips. After the dried CNF surfaces being imaged by AFM, the relative nanosize could be verified in AFM image of Figure 1(f), which showing rod-shaped crystals (100–200 nm long by 5–10 nm wide). The dried CNF were packed in sealed plastic bag and stored in a refrigerator at 5-10°C prior to melt-compounding.

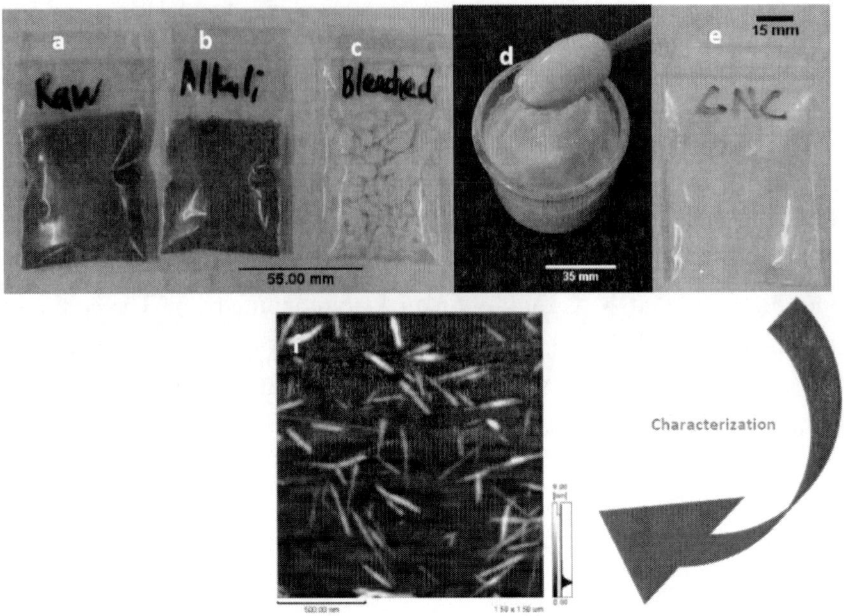

Figure 1. Photographs of (a) the ground raw kenaf fiber, (b) Alkali-treated kenaf fibers, (c) bleached kenaf fibers, (d) gel form acid-hydrolyzed-fiber, (e) Freeze-dried cellulose nanofibers (also known as cellulose nanocrystal, CNC) and (f) AFM images of cellulose nanofiber as referred to Santos et al. (2013) [15].

2.3. Preparation of TPS and Nanocomposite Films

The preparation of TPS and nanocomposite films according to formulations of Table 1 was based on consecutive steps of twin-screw extrusion and compression moulding. The plasticized starch was merely prepared by manually blending of the indicated starch, water and glycerol and stearic acid. To ensure the homogeneous redispersion of freeze dried CNF (aggregate form) in matrix, the varied amount of CNF was fully dissolved into the mixture of plasticizers (deionized water, glycerol) first prior to pouring into mixture of starch and stearic acid. The Na-MMT is the last material dried mixed with plasticized starch with or without CNF. The plasticized starch were then fed into twin screw extruder (Giant Instrument Manufacturing Co., Ltd., Nanjing, China) with a screw speed of

360 rpm and the temperature profile was varied from 105°C at the feeding zone to 120°C at the die, synthesizing CONTROL films. For prepared premixes containing thermally stable nanofiller, their inherent high heat resistance might retard the heat-induced starch gelatinization, thus their corresponding temperature profile was varied from 110°C at the feeding zone to 125°C at the die. Afterwards, the cut ribbon extrudates were filled in a metal mold (150mm of width and length, 1mm/3mm of thickness) followed by subjecting to hot-press machine (Collin PCS-GA Type Press 800) for 3 min preheating and 5 min hot-pressing at 125°C under a pressure of 10 MPa. For testing purpose, the dumb-bell shaped specimens prepared from (1 mm thick) hot-pressed films while notched impact test bars prepared from 3 mm hot-pressed sheets, they were conditioned in a desiccator over silica gel prior to non-ageing characterization or thermal–ageing.

Table 1. Formulations of starch-based compounds

Sample code	**Loading Level (phr[a])**					
	Starch	**Glycerol**	**Water**	**Stearic Acid**	**CNF**	**MMT**
TPS[b]	100	32	16	4	0	0
0C10M	100	32	16	4	0	10
2C8M	100	32	16	4	2	8
4C6M	100	32	16	4	4	6
6C4M	100	32	16	4	6	4
8C2M	100	32	16	4	8	2
10C0M	100	32	16	4	10	0

[a] parts per hundred parts of starch.
[b] CONTROL.

2.4. Thermal Ageing Studies

In order to evaluate the durability of the used TPS and nanocomposites, an accelerated thermal-oxidative treatment procedure, that simulates severe conditions for thermal aging, was adopted. The prepared samples were stored in air of the forced-air circulation oven (ESPEC facilities, Japan) set at temperature of 60°C, the choice of this temperature is essentially justified by the simulation of severe aging conditions where the service temperature highly probable to reach. The specimens with dumbbell-shape and impact test bar were aged for tensile test and impact test respectively. An accurately weighed amount of samples (tensile specimen + impact test bar) for each formulation was subjected to isothermal ageing by six individual sets of samples with equal quantities, they were simultaneously subjected to ageing experiment of 6, 12, 18, 24, 48 and 72 hours heating. At various stages of ageing, the heat-treated samples of all formulations were were taken out, air cooled in a desiccator and characterized. The keeping of aged samples in desiccator was also to eliminate possible residual thermal effects and also to avoid the complicating influence of water absorption prior to characterization. The thermal ageing experiment was maximally lasting for 3 days at 60°C considering the specimens became too brittle for handling or even tensile testing after being exposed to additional ageing time.

2.5. Characterizations

2.5.1. Tensile Test

Dumbbell specimens were cut out from the compression moulded sheets according to the ASTM D638 (type IV) and the gauge thickness, width and length of non-aged and microwave-aged samples were measured prior to tensile testing. The uniaxial tensile test was carried out at laboratory temperature using an Instron Universal Testing Machine (Model 4302 Series IX) equipped with a 2 kN load cell and with a constant crosshead speed of 50 mm/min. The final value of each tensile properties data was calculated from the average of 5 specimens of each sample group.

2.5.2. Scanning Electron Microscope (SEM) Analysis

The fractured surface of samples after tensile failure was morphologically observed by a scanning electron microscope (SEM) using the model of Hitachi Scanning Electronic Microscopy of BS 340 TESLA. Prior to scanning, the samples were cut into a smaller portion and then mounted onto the copper stub using double-sided tapes with the fractured surface was facing up. These samples were first coated with a thin layer of gold for conduction of electron to avoid any disturbance of electron beam to the molecular structure and just can be available for imaging. The scanning electron micro photographs were recorded at a magnification power of 3000X which was operated at the electron beam voltage of 20 to 30 KV.

2.5.3. Water Absorption Test

Water uptake measurement was performed using the tail end-parts of dumbbell specimens with 9 × 18 × 1 mm dimensions (width x length x thickness). Before exposing to water, the unaged/aged samples cuboids were weighed for the original dry weight (Wo). After soaking in distilled water for 10 min at room temperature, the test pieces were then removed from distilled water and gently dried by using air blow gun, and immediately weighed to obtain the weight (W_1). Here, the capability of water absorption was expressed by Equation (1):

$$\%water\ absorption = \frac{W_1 - W_o}{W_o} \times 100\% \tag{1}$$

The measurements for each sample were performed in triplicates.

2.5.4. Impact Test

Izod impact resistance was determined according to the technical standard method ASTM D256, using pendulum impact tester. The sample bars with a V-shape notch for notched Izod impact test were cut from the 3 mm thick sample sheets while each of the test bar has standard dimensions of 12.7 × 63.5 × 3 mm^3 (width×length×depth), following by making 10.16 ± 0.05 mm indentations in the test specimens. All the measurement of notched Izod impact strength was carried out at 23 ± 2°C and 50 ± 5% relative humidity. The obtained impact strength value of each sample was calculated according to Equation (2) by taking the average of five values. In the case of COV (coefficient of variance) exceeded 5%, the number of tested samples would be increased.

$$\text{Impact strength} = \frac{W}{h \times b} \times 10 kJ/m^2 \tag{2}$$

while W = specimen impact broken energy (J)
h = specimen thickness (mm)
b = specimen width (mm), from indented point to lateral surface

3. Results and Discussion

3.1. Mechanical Properties Analysis

Figure 2 show the effect of high temperature exposure on the mechanical properties of TPS and nanocomposites as a function of ageing time, while the tensile strength (TS) and Young's modulus (E) before and after thermal ageing could be compared in Figure 2 a-b excluding the % elongation at break (EB) as shown in Figure 2c considering the EB value of non-aged samples too large to compare. Here, the obtained EB value of non-aged samples is recorded as 115.939 ± 6.26 (TPS), 90.806 ± 15.034 (0C10M), 77.837 ± 10.727 (2C8M), 52.965 ± 9.455 (4C6M), 72.725 ± 6.807 (6C4M), 83.88 ± 7.822 (8C2M), 38.834 ± 17 (10C0M).

The TS of composite materials mainly depends on the efficiency of stress transfer from matrix to filler [27]. The E of polymer represents the ability (restriction of chains' mobility) or stiffness to resist the mechanical-deformation retarding the material from losing its elastic properties. The stress transfer and restriction of chains movement mainly relying on the intermolecular interaction among components, while EB is reciprocal to this interaction which tends to limit the sliding of inter-molecular chains. By simultaneously incorporating CNF/MMT into starch, the hydrophilic constituents can interact in different manners with different degree of intermolecular interaction depending on their competency/dynamic to develop hydrogen bonding at their interface. The final expressions about the relevant hydrogen bonding interaction formed at the interface between hydrophilic components in studied nanocomposites have been proposed in Table 2.

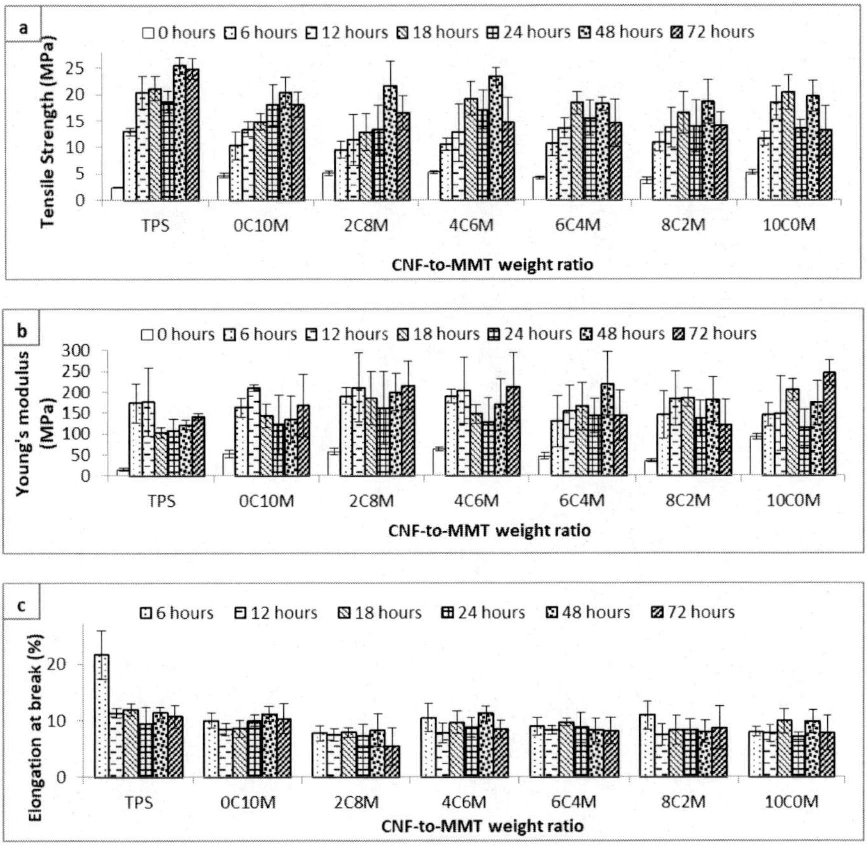

Figure 2. (a) Tensile strength, (b) Young's modulus and (c) percentage elongation at break of thermal aged TPS and nanocomposites with various CNF-to-MMT ratio as a function of ageing times.

Table 2. Proposed hydrogen bonding interactions among hydrophilic components within nanocomposites

Formulations	Representative hydrogen bonding interactions
0C10M	Starch-OH+HO-MMT
2C8M	(Starch-OH + HO-MMT) + Cellulose-OH (minor) ⟷ MMT-OH + HO-Starch-OH + HO-Cellulose (major)
4C6M	MMT-OH + HO-Starch-OH + HO-Cellulose ↕ Starch-OH+ HO-(Cellulose-OH+HO-MMT)-OH+HO-Starch
6C4M	MMT-OH + HO-Starch-OH + HO-Cellulose (minor) ⟷ (Starch-OH + HO-Cellulose) + MMT-OH (major)
8C2M	(Starch-OH + HO-Cellulose) + MMT-OH
10C0M	Starch-OH + HO-Cellulose

⟷ Represents competency.

Under non-ageing condition, the plasticizing effect of glycerol and water conferred TPS with high flexibility but low strength and stiffness before being reinforced by nanofiller as shown in Figure 2. After introducing nanofiller, the intermolecular bondings formed at the matrix-nanofiller interface could allow the effective stress transfer between matrix-filler and restrict the movement of starch chains when being subjected to stress straining, causing the higher TS and E but lower EB of nanocomposite than non-filled TPS [27]. The improvement of TS/E of TPS by the addition of 10 phr MMT mainly attributed to the interaction of starch with one edge hydroxyl group of MMT and intercalation system of starch chains within silicate gallaries. The intercalation of polysaccharides chains into MMT layers galleries was possibly occurred in the studied bionanocomposites since single polysaccharide chain is a linear molecule with monomer size of about 0.55 nm which is lower than d-spacing difference between pristine clay layers, allowing strong polar interaction of starch-MMT or CNF-MMT [16]. With partially substituting MMT with 2-4 phr CNF, the mechanical properties were desirably improved continuously from 0C10M to 4C6M. By similar polysaccharides structure, it is believed that the higher similarity of surface chemistry between CNF-starch tend to create higher degree of intermolecular hydrogen interaction at the multi-interface in 2C8M and 4C6M than 0C10M. Meanwhile, this positive effect could be amplified by higher volume fraction of CNF (since $\rho_{CNF} < \rho_{MMT}$) and decreased MMT weight fraction, which contributing to more reactive sites could be totally provided by filler for developing intermolecular bonding. From another point of view, the increased total amount of polysaccharides chains (starch+CNF) and decreasing amount clay galleries from 0C10M-4C6M has considerably allowed the extensive diffusion of starch-CNF to push the platelets far apart till the initiation of exfoliation mechanism [13, 28, 29]. Under the dominant MMT dispersion in 2C10M-4C6M, the exfoliation system of polymer chains among silicate layers is well-known to improve the strength/rigidity of nanocomposites. With further increasing CNF-to-MMT ratio continuously, the increasing trend of exfoliation mechanism might have been stopped at 6C4M-8C2M as observed from relevant reduced TS/E. Under the dominant CNF population in matrix of 6C4M-8C2M, it is believed that the higher dynamic of starch to form hydrogen bonding with CNF by their higher chemical similarity tended to individualize MMT as small particles beyond starch-CNF interface. The worse is such small particles might act as defects among dominant starch-CNF matrix to probably interfere formation of intermolecular hydrogen bonding in 6C4M-8C2M. This statement could be supported by much higher TS/E of 10C0M than 6C4M-8C2M indicating the large difference in their intermolecular bonding force, thus excluding the initiation of intercalation mechanism in 6C4M-8C2M unless otherwise MMT defects formation.

Regarding to thermal-aged samples, the observed large error bars on the results of TS/E indicate that there may be the large difference in structural properties varied from surface till core of the samples evolved by heat stresses and auto-oxidation. Apparently, the overall strength and stiffness of aged samples is much higher than non-aged samples over the ageing time, but the poorer strain at break has been residual effect of thermal ageing. On this basis, it can be deduced that the cross-linking-consolidation effects overrides the degradation mechanism over the ageing test. The initial improvement of material rigidity from 0-6 hours ageing mainly comes from the loss of volatile matter involving water molecules or tightly bound water and plasticizer present in the starch molecules [5]. The high temperature exposure lasting for 6 hours had an especially greater influence on the TS/E of on non-filled TPS than nanocomposite, implying higher heat ageing resistance of nanofiller-added composites. The high heat resistance could be contributed by high stability of nanofiller and/or intermolecular bonding between filler-matrix nanofiller to heat flow, towards retarding the heat gaining for evaporation of volatile matter. The nanoparticles also possibly interacted with volatile decomposition products to retard their diffusion out of the sample [30]. Apart from the absence of nanofiller, branched structure (along amylopectin) of starch is another determining factor to facilitate the formation of starch crosslinking networks under heat treatment [22]. The reduced EB during 6-hours ageing stage mainly due to the reduced plasticity of material by loss of volatile matter, one should mention that the higher ductility of pure TPS than nanocomposites could be ascribed to the lack of nucleating effects of nanofillers to induce thermal-recrystallization process which increased the brittleness [28].

As a function of ageing time, the further improve in mechanical properties from 6 to 18 hours mainly resulted from the crosslinking of free radicals generated by thermal-dissociation of C-H bonds along polysaccharides chains [22]. Apart from C-C crosslink, C-O-C is another possible crosslink via ether and/or ester linkages between hydroxyl (-OH) groups of the hydrophilic constituents since intra- and intermolecular bonding have been recovered upon water or plasticizer within starch inter-chains was evaporated. The formation of crosslinking (especially C-O-C) was considerably accelerated by the faster volatile evaporation rate in TPS and 6C4M-8C2M previously to provide free volume for crosslinking, mainly due to their weaker intermolecular bonding with lower thermal resistance to retard evaporation. This also prove that the CNF could more suitably serve as post-cure agent than MMT by emanating free radicals or providing –OH groups for crosslinking. The increment of TS/E and decrement of EB with increasing ageing time mainly attributed to the increased crosslink density with higher molecular weight enabling greater efficiency of stress transfer and the increase in the molecular entangling via the post-crosslinking to reduce the molecular mobility of starch chains [1, 11, 19]. However, there is insignificant increase of TS of 0C10M-2C8M, it is reasonable to state that the presence of high amount silica which retarded the crosslinking by trapping radicals from reaction with its larger surface area (~800 $m^2 \cdot g^{-1}$) than CNF (several hundreds of $m^2 \cdot g^{-1}$) [30,31]. Moreover, there are limited gallery spaces within silicate galleries available for crosslinking [32]. As compared to 0C10M-4C6M, the 6C4M-10C0M has insignificant increment of tensile modulus during 6-12 hours ageing stages. This behavior is related to the restricted movement of the polymer chains/macromolecules in the vicinity of the cellulose nanofibers via strong hydrogen bonding for recrystallization of semi-crystalline TPS [31, 33]. It is well-known that the increased crystallinity of polymer could bring in increase of the tensile modulus of the nanocomposites whereas crystallization behavior is sensitive to heat [35]. The significant increase of

E of 0C10M-4C6M probably related to the weaker intermolecular bonding of starch with MMT than with CNF, enabling more mobile units participated in the isothermal-recrystallization through clay galleries guided by layer effect of MMT [1, 12]. However, the E of 0C10M-4C6M couldn't be increased continuously at 18 hours ageing instead of the decrement of E, suggesting misalignment of polymer chains rather than thermal-degradation on crystalline structure, which imposing adverse effect on the crystallinity and even tensile modulus [30, 33]. Such misalignment of polymer chains might be induced by improper nucleating action of silicate layers unlike increment of E in 6C4M-10C0M upon gaining sufficient heat energy for rendering isothermal-recrystallization [33, 36].

Further prolonged ageing (18-24-48 hours), there were descending and ascending of TS/E for TPS and nanocomposites likely due to the decreased crosslinking density followed by molecular rearrangement of remainders. The crosslinking density considerably reduced by degradation mechanism initiated at 24 hours ageing through chain scission which might occur in many positions in the crosslinking network and amorphous/crystalline region, while thermolysis of bridge between crosslinking points and polymer chains can be expected [20]. Then they escape from the thermolysis system in the form of small molecules, molecular rearrangements may occur among the remainders upon continued ageing to 48 hours, resulting in the formation of more stable compound with an increased crosslink density [20]. Since there was lack of formation of crosslinking in 0C10M-2C8M during initial ageing, the decreased crosslinking density can be excluded at 18-24-48 hours ageing stage instead of increased degree of crosslinking as realized by continuous increment of TS. This phenomenon was mainly related to increasingly stronger post curing effect could be induced by prolonged heat exposure towards achieving optimum cured samples unlike slower rate of silicate layers inducing saturation of crosslinking previously [11]. Thus, the further misalignment of polymer chains followed by recovery of recrystallization or increased crosslinking density till saturation can be a possible explanation for descending and ascending of E in 0C10M-2C8M during 18-24-48 ageing stage.

In the final stage (48-72 hours), the deterioration of TS can be induced by accumulative thermal-oxidation till the increased thickness of oxidized layer, leading to the uneven stress distribution or discontinuous load transfer since there were much lower load available to be transferred in oxidized region which comprised of ROO• and RO•/HO• decomposed from ROOH [20, 33, 37]. The insignificant reduction of TS in 0C10M might be due to thermal-oxidation occurred in line with crosslinking formation but oxidative-degradation still prevail over crosslinking. The surface damages by oxidative degradation and increased crystalline structure led to samples exhibiting brittle behavior, explaining the reduced extensibility (EB) at 72-hours ageing stage [19]. Since crystalline structure was considerably impermeable to oxygen, the E of samples was unaffected by oxidative degradation instead of crystallinity and restriction of molecular motion being probably improved by recrystallization and continuous molecular recombination of remainders. However, the E increment was excluded for 72-hours-aged 6C4M-8C2M, likely due to the optimum crystallization has been reached at 48 hours followed by misalignment of polymer chains or the disability of CNF's nucleation to sustain till 72 hours ageing under the disturbance of MMT defects. Notably, this situation couldn't be explained by reason of crystalline chain-scissioning as realized by good improvement of E in 72-hours-aged 10C0M likely through good nucleating ability of highly crystalline CNF in providing primary nucleation sites for geometry growth of crystals [19, 36].

3.2. Morphological Structure Analysis

In order to confirm the mechanical modifications previously discerned, cross-sectional observations of the studied materials have been investigated using scanning electron microscopy (SEM) [34]. Figures 3 to 5 demonstrate the fractured surface observations of TPS and nanocomposite with various CNF-to-MMT ratios under condition of 0 hours vs 6 hours, 18 vs 24 hours and 72 hours thermal ageing respectively.

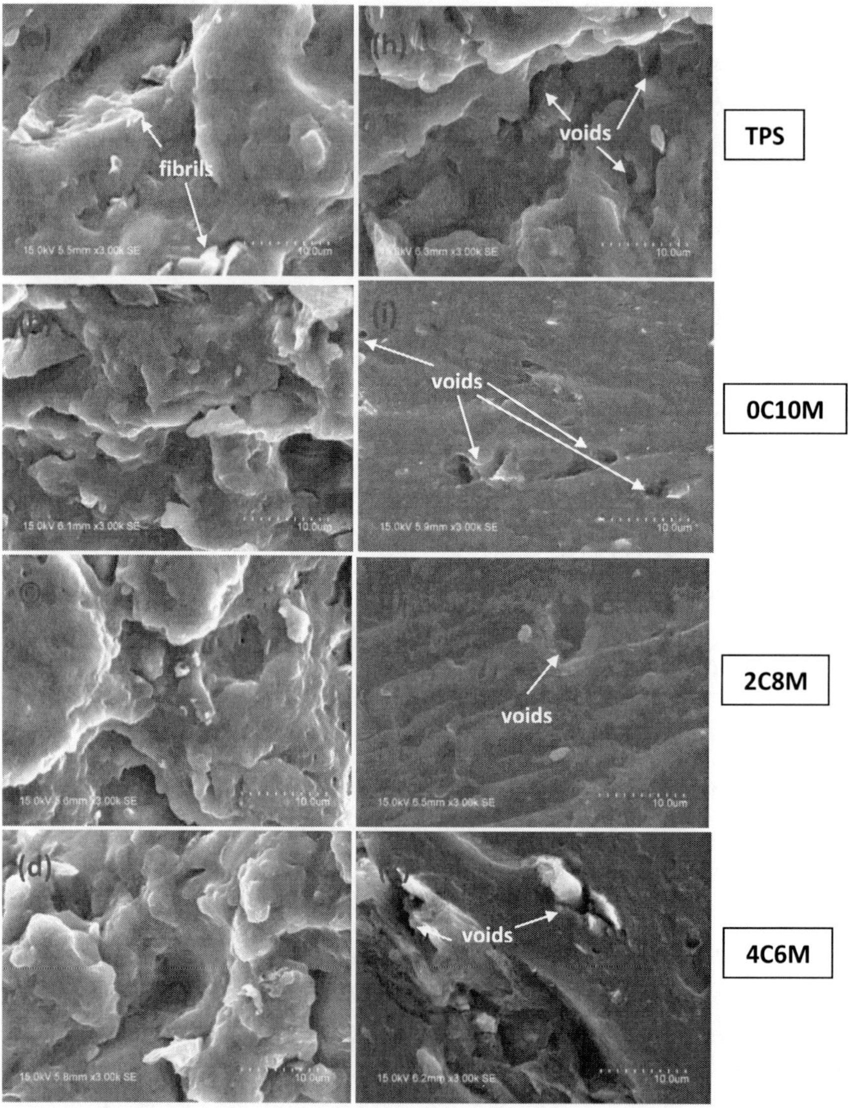

Figure 3. (Continued).

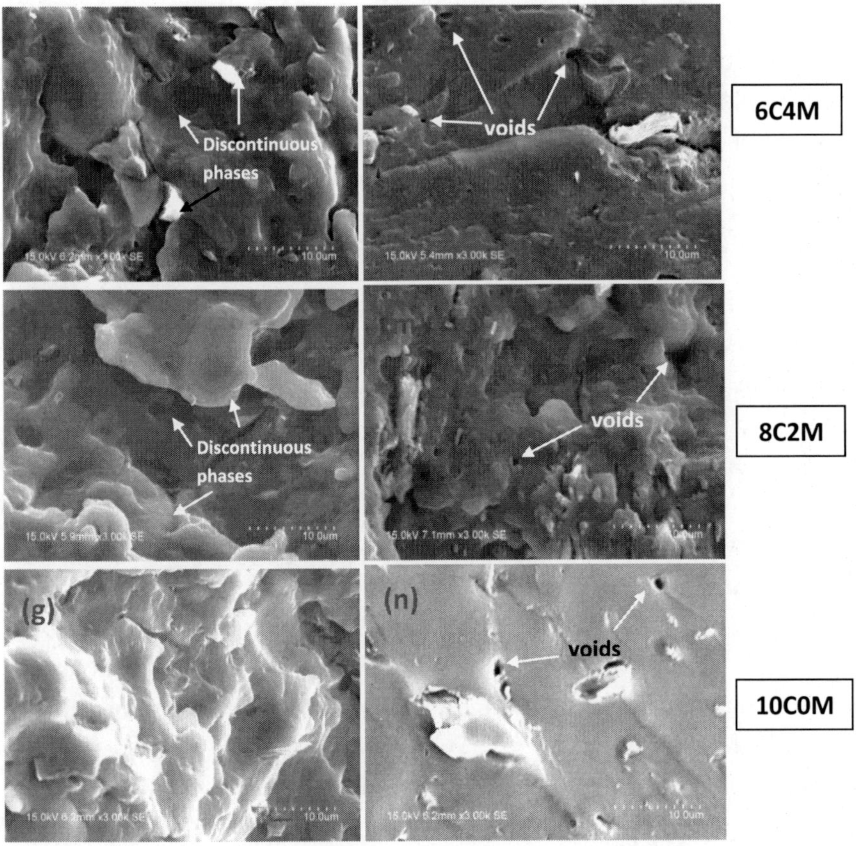

Figure 3. SEM micrograph at 3000X magnification of fractured surface of 0 hours (a-g) and 6 hours (h-n) thermal-aged TPS and starch-nanocomposite filled CNF/MMT with weight ratio of 0/10, 2/8, 4/6, 6/4, 8/2 and 10/0 respectively.

As demonstrated in Figure 3, the thermal ageing has imposed significant effect on fracture surfaces of non-aged samples. In contrast to non-aged film samples (Figures 3a-g) of having smooth surfaces without cracks visible to the naked eye, Figures 3h-n show a series of morphology changes due to thermal aging in which the crackles structure were formed [33,34]. This is an indication of the modification of the fracture mode from ductile to brittle. The characteristic of ductile failure can be evidenced by wrinkles formed by tearing the matrix into fibrils during tensile straining, indicating high flexibility of non-aged matrix [34]. One can compare that the fibrils on pure TPS become shorter upon the addition of 10 phr MMT as shown in Figure 3b considering the nanoreinforcements would cause stress concentrations to reduce the extensibility of matrix [29]. For 6 hours aged samples, the reduced formation or loss of fibrils indicating the deterioration of high plastic deformation prior to fracture and thus decreased plasticity (strain) and even increased brittle/crackle behavior. In case of aged TPS, the fibrils still can be clearly seen after ageing, evidencing its good sustainability of plasticity without nanofiller. The formation of pores/voids is another feature of showing the dramatic change in the fracture topography by 6-hours ageing [33, 34]. This observation suggests the loss of material from near the core of the heat-treated matrix, probably being correlated to the evolution of materials-volatile or otherwise- from within the matrix, via the surface [33]. More voids could be found in because of formation of discontinuous phases as shown in Figure 3e-f was more prone to releasing of volatile matter. Unlike the poor continuity of matrix phases of 6C4M-8C2M caused by isolating effect of MMT, the 2C8M-4C6M have better continuity mainly due to the partial substitution of MMT by CNF could develop stronger intermolecular hydrogen interaction between nanofiller-starch to reduce discontinuity of matrix phase [38, 39].

The fractured surfaces captured in Figure 4 reveal the evolution of the surface morphology behavior from 18 to 24 hours ageing. As compared to 6 hours-aged films, such fracture surface turned to be rougher at 18 hours ageing in which the sharper projections/edges could be observed, implying a more brittle failure. The increased crosslinking-density might be reasonable to explain such increased surface roughness of fractured surface that is a result of strong deformation resistance during stretching by high matrix strength from 3D crosslinking network, while this strong yielding effect considerably causing increased multiple crack sites [38, 39]. In addition to the rougher surface of crosslinking, the matrix phases of 0C10M-4C6M (Figures 4b-d) also displayed higher level of discontinuity till 24 hours ageing (Figures 4i-k) as a consequence of misalignment of polymer chains upon optimum recrystallization was reached.

In a dissimilar manner, the 18-24-hours aged samples with higher CNF fractions (6C4M-10C0M) displayed greater orderliness of tear-lines as can be found in Figure 4e-g and 4l-n, suggesting the highly ordered structure has been stabilized by great nucleation potential of CNF irrespective of decreased crosslinking density at 24 hours ageing. As evidenced by Figures h-n, such reduced crosslinking density by chain scissioning could only affect the fracture face of the composite which turned to be smoother from high roughness. This behavior indicates major damage of the crosslinked starch matrix by thermolysis leading to the scissioned chains become more susceptible to phase separation by applied straining stress without any significant deformation resistance due to insufficient matrix stiffness [16]. Such smooth morphology could be except for the samples with higher weight fraction nanoclay (0C10M-4C6M), since surface roughness (or sharp projections) in the surface of the films were aggravated by the tardily increased crosslinking density [16].

For 72-hours-aged samples, Figure 5 showing that the starch films and starch/nanofiller polymer blends had a relatively smooth and continuous matrix with low valleys and hills and with good structural integrity. The matrix was so degraded chemically in the oxidized region to such an extent thereby leading to the behavior of the composite in the oxidized region was practically that of a facile separation of phases without prominent resistance [19]. For the insignificant surface damage in CNF-filled samples, cellulosic filler itself might act as an antioxidant agent with hydrogen-donor or a scavenger of free oxygen radicals to terminate the chain reactions during the autooxidation process like production of hydroperoxides and even RO•/HO•, reducing the degree of oxidation [21, 37]. Since there were insignificantly oxidative damage on crystalline region, high continuity of phases or orientated tear-lines still can be clearly seen on 72-aged matrix excluding the 6C4M-8C2M. As shown in Figure 5d-e, the disordered/irregular tear-lines seems to be well correlated to the abnormal recrystallization behavior under disturbance of MMT defects on thermal-rearrangement of polymer chains.

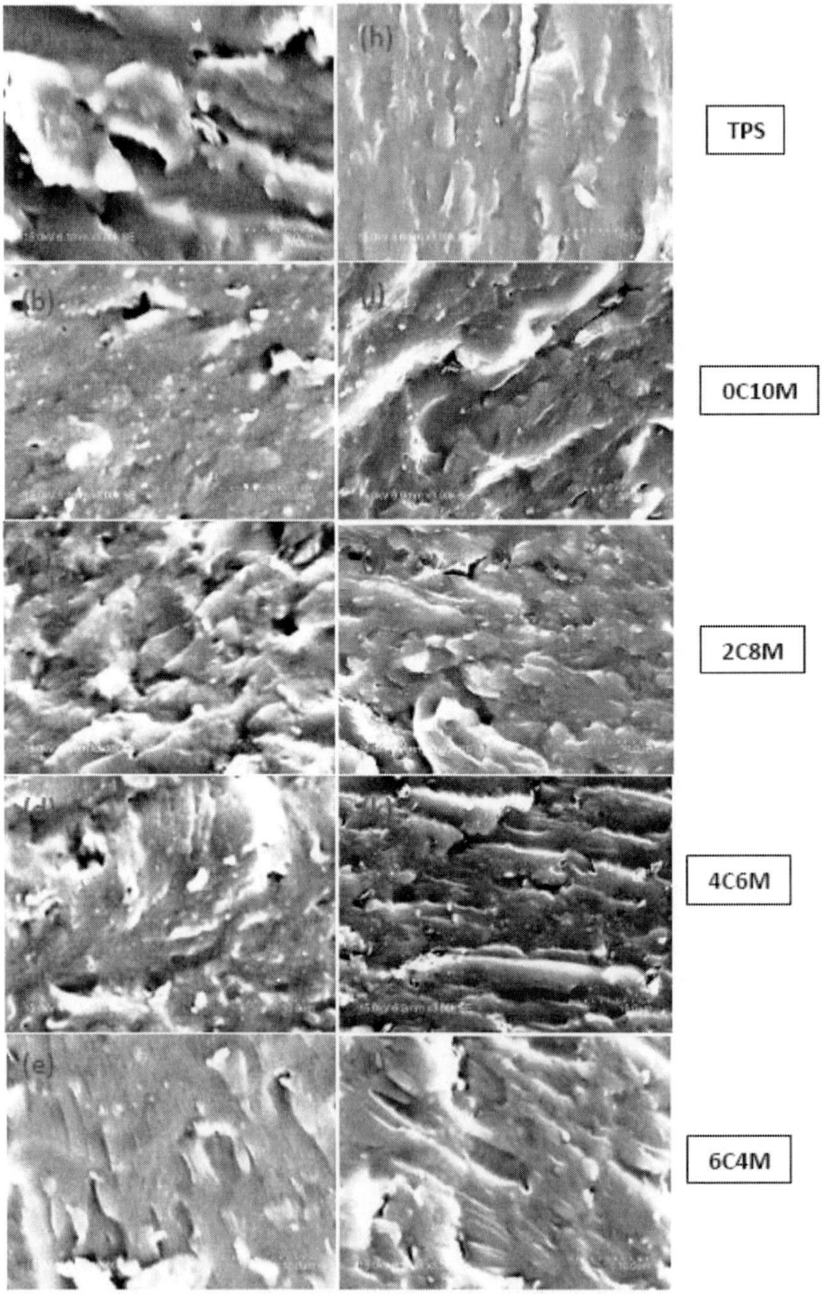

Figure 4. (Continued).

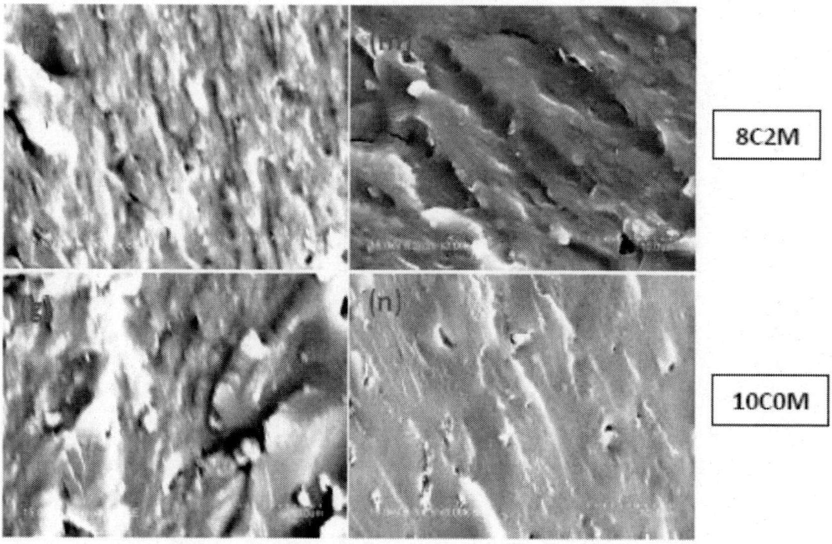

Figure 4. SEM micrograph at 3000X magnification of fractured surface of 18 hours (a-g) and 24 hours (h-n) thermal-aged TPS and starch-nanocomposite filled CNF/MMT with weight ratio of 0/10, 2/8, 4/6, 6/4, 8/2 and 10/0 respectively.

3.3. Water Absorption Properties

Water absorption was an important index to estimate the utility of starch-composites since water likely impart plasticizing effect on starch matrix and also induce the growth of microorganisms, such as bacteria and fungi, access to the interior of the matrix [40, 41]. Concerning to the possibility of studied starchy composites being prone to microorganism attack as well as biodegradation through enzymatic cleavage on hydrolysable backbones, the water resistance was investigated by referring to plotting graph (Figure 6) of percentage water absorption of TPS and nanocomposite as a function of ageing time from 0 to 72 hours.

Since starch molecule bearing abundant hydrophilic functional groups (-OH), the highest water absorption of neat TPS was expectedly obtained among non-aged samples. However, the unfavorably high hydrophilicity of TPS would be reduced after being incorporated by nanofiller. Response to similarly high water resistance of 0C10M and 10C0M, MMT and CNF itself alone considerably already possesses ability to restrain the penetration of water by the clay of making surface more hydrophobe whereas the CNF could reduce the hydrophilic groups (-OH) exposed on TPS through extensive hydrogen bonding [42, 43, 44]. In case of simultaneous addition of MMT/CNF, the additional factor decreasing the water-sensitivity of TPS mainly belonged to the formation of comparably higher degree of intermolecular hydrogen bonding at multi-interfaces between MMT-starch-CNF. The resulting stiff intermolecular interface could serve as strong water barrier, contributing to the lower %water absorption of 2C8M and 4C6M than 0C10M. Due to the existence of MMT defects within the 6C4M and 8C2M, the gaps might be exploited along the interface between dominant starch-CNF phase and isolated MMT, such diffusion pathway tend to allow water entrance into the composites system [45]. Moreover, the crystallization process also possibly disturbed by MMT defects resulting in less crystalline region could be formed considering compact system of crystalline structure is impermeable to water. As a result, 6C4M and 8C2M possessed lowest water resistance among non-aged nanocomposites.

As long as the thermal ageing was initiated until 6 hours, the micro-voids left behind by loss of volatile matter possible to allow the water penetration, explaining the increased water absorption of most 6-hours aged samples. In case of improved water resistance in 6-hours aged TPS and insignificant increase of water absorption in 6C4M-8C2M, the sufficient degree crosslinking network was considerably achieved in faster rate by their weaker heat ageing resistance and faster evaporation of volatile material, leading to more efficiently resist water entrance. It is note-worthy that the formation of three-dimensional cross-linking network structure is interconnected polymer chains with high rigidity which are insoluble in water and infusible to water [1]. Due to evaporation of volatile matter highly restricted by extraordinarily stable intermolecular bonding in non-aged 4C6M and 10C0M, the comparably higher water absorption was resulted since the poorer crosslinking (by slower liberation of C•) and less micro-voids have been formed.

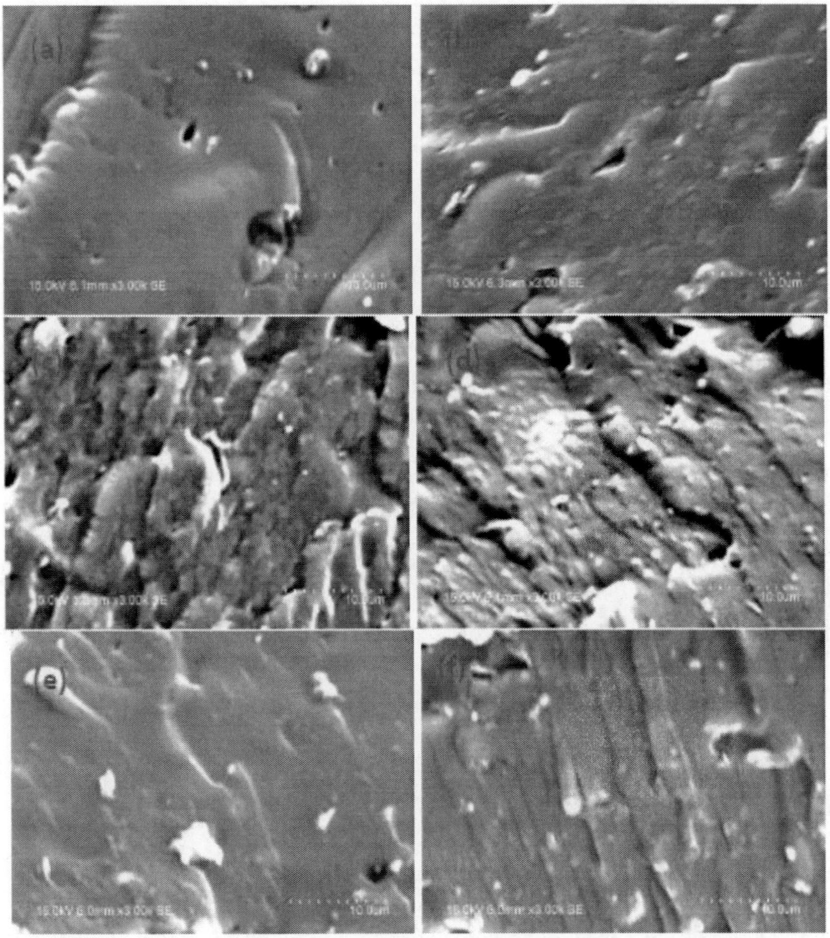

Figure 5. SEM micrograph at 3000X magnification of fractured surface of 72 hours-aged (a) TPS and starch-nanocomposite filled CNF/MMT with weight ratio of (b) 0/10, (c) 4/6, (d) 6/4, (e) 8/2 and (f) 10/0 respectively.

Since crosslinking could be increasingly formed with ageing time, the increased cross-linking density consequently improved the water resistance of 6-18 hours-aged samples especially 4C6M-10C0M. By comparison of this ageing stage, CNF showed better potential as post-cure agent than MMT clay to overcome the high hygroscopic property of aged starch film by forming higher intensity of crosslinking [38]. This can be supported by the smaller swelling ratio represents the higher degree of cross-linking network has been obtained since water swelling behavior is one of the most important structural parameters characterizing a cross-linked polymer [1]. However, the strong reduction of water absorption is excluded in 12-18-hours aged 0C10M-2C8M, indicating the insufficient formation of crosslinking achieved by less free radical (mostly blocked by silicate layers) participated in crosslinking. Thus, the slight improvement of water resistance in 12-hours aged 0C10M-2C8M can be possibly ascribed to isothermal-recrystallization-increased crystalline structures which considerably improved water impermeability. In turn of efficiency of water resistance marginally decreased in 18-hours-aged 0C10M-2C8M, suggesting the misalignment tendency of MMT imposed on starch chains by chemistry dissimilarity between MMT-polysaccharides could alter the inherent crystallite structure as well as reduce water barrier. In a different manner, the water resistance could be increased in 18-hours-aged 6C10M-10C0M likely due to the good nucleation ability of CNF to increase water-impermeable crystalline region.

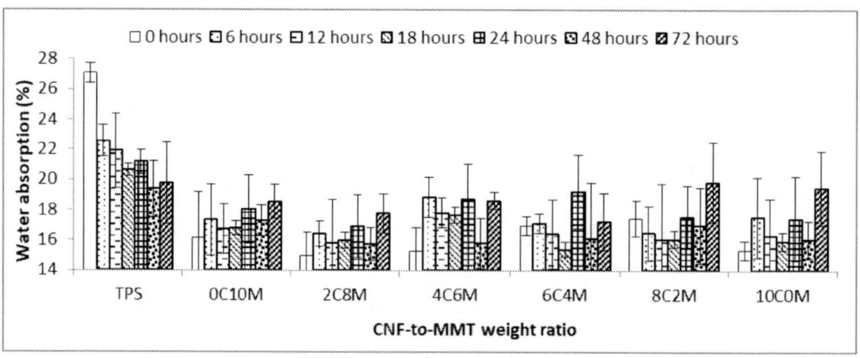

Figure 6. Percentage water absorption of thermal aged TPS and nanocomposites with various CNF-to-MMT ratio as a function of ageing times from 0 to 72 hours.

Within the prolonged ageing time (24-48 hours), the magnitude of water absorption increased firstly followed by decrement. The enhanced water diffusion mainly related to the free volumes/interspaces provided from chain-scissioning or thermolysis of crosslinking bridges that were available for increasing water holding capacity. Apart from detrimental effect of thermal degradation, thermal expansion of aged samples which could increase the water swelling volume also possible to allow more water absorption. The water resistance was recovered at 48 hours ageing by rearrangement of molecular remainders or further reorientation of misaligned polymer chains. At ageing time of 72 hours, the thermal-oxidative time in excess has increased thickness of the oxide layer leading to the water resistance declined by its loosening structure which resulted from the dissociation of ROOH or R-H bonding [20].

3.4. Impact Resistance

Impact strength is the ability of a material to resist fracture under stress applied at high speed [46]. Fracture toughness and impact resistance are among the most important material's property requirements. Determination of fracture characteristics or understanding of the fracture initiation and propagation process, therefore, is extremely important especially for application in household food container [47].

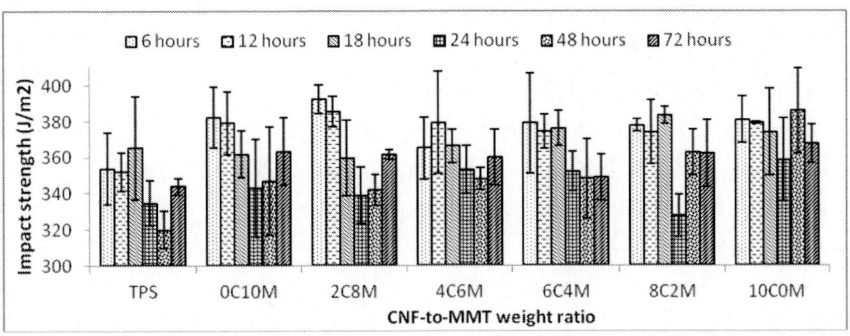

Figure 7. Izod impact strength of thermal aged TPS and nanocomposites with various CNF-to-MMT ratio as a function of ageing times from 0 to 72 hours.

Figure 7 shows the notched Izod impact strengths of thermal-aged TPS and nanocomposites at different CNF-to-MMT ratio, while the impact strength values of non-aged samples are regarded as 767.93 ± 64.47 (TPS), 616.15 ± 17.854 (0C10M), 491.916 ± 33.07 (2C8M), 486.953 ± 70.41 (4C6M), 659.104 ± 45.094 (6C4M), 602.207 ± 79.995 (8C2M) and 647.29 ± 37.7 (10C0M) J/m^2.

Prior to thermal ageing, the Izod impact results show that the non-aged specimens have a significantly higher capacity of absorbing impact energy implying their high plastic deformability with good flexibility [48]. In the absence of nanofiller, the highest impact strength of TPS among the non-aged samples could be observed. Without filler-matrix interface which is favorable to initiate fracture by debonding, the neat TPS possessed good fracture initiation resistance, thus highest impact loading required for deformation of matrix polymer and even impact fracture [47]. After the inclusion of nanofiller, impact properties of nanocomposite were affected by fracture initiation and crack propagation mechanism due to the presence of interface between the filler and the matrix polymer could constitute the stress concentrating points for inducing initiation of fracture [47]. It is therefore less impact energy necessary to be absorbed for initiating debonding of filler-matrix interface during the fracture process of nanocomposites and lower impact strength than TPS consequently [47]. Upon substituting MMT by higher volume fraction CNF, the decreased impact strength in 2C8M-4C6M could be explained by increased quantities of debonding at matrix/reinforcement interfaces since higher volumetric loads of CNF than MMT since $\rho_{CNF} < \rho_{MMT}$ [47]. Implicitly, the increased crack initiating points enabled to easier cracking propagation by lower impact strength. From the view point of higher impact resistance of 0C10M than 2C8M-4C6M, suggesting the compact intermolecular structure of intercalation system had poorer stress concentrating ability to induce debonding mechanism and lower cavitation/voiding effect could be induced for cracking propagation [47]. Otherwise, the higher tendency of exfoliation mechanism occurred in 2C8M-4C6M with further apart distance between silicates layers enabling cracks to be easily propagated by filler/matrix debonding near free surface, resulting in another factor of decreased impact strength [47]. Favorably, further substitution of MMT with CNF from 6C4M until 8C2M could increase the impact strength under the dominant distribution of much higher amount CNF within starch matrix. This ideal result can be contributed by the MMT with minor portion might act as lubricant to exert plasticizing effects on dominant CNF-starch phase, the subsequently increased plasticization therefore able to effectively absorb high speed impact to higher extent before impact fracture. By taking 8C2M as comparison, 10C0M had no such MMT defects instead of good CNF-starch adhesion by relevant similar polysaccharides structure, thus contributing to increased impact strength by an effective resistance to fracture initiation and crack propagation [46, 47].

In comparison with data of non-aged notched samples, the sharp decrement of impact strength after applying high temperature severe ageing can be globally observed in Figure 7. Since there are showing no big difference in impact strength values irrespective of variation in ageing time and composition, the reason of large drop of impact strength to approximate level around 350 J/m^2 without significantly large error bars can be simply classified into embrittlement effect. One can assume such embrittlement can be caused by homogeneous consolidation/degradation mechanisms throughout thickness of the material, thermal-recrystallization, and the last is surface-oxidation effect [19]. In the initial ageing (6-18 hours), the generating consolidation effect mainly resulted from evaporation of volatile matter and formation of crosslinking [48]. The loss of water and plasticizer by heat tended to reduce the intrinsic plasticization behavior or plastic deformability of those blends, representing the weakened ability to absorb impact energy for fracture initiation or reduced matrix polymer's resistance to crack propagation [47]. The formation of crosslinking leads to consequently representative behavior of the mechanical properties along with an increase in the hard elastic network typical of brittle material [48]. If all impact energy was consumed to propagate fracture for crosslinked-notched samples, this energy was considerably very low implying just brittle fracture would be occurred without significant resistance during cracking propagation in the boundary condition [47]. The high brittleness also possibly induced by increased crystallinity from recrystallization process, contributing to lower impact resistance by increased cracking sensitivity mostly occurred in 0C10M-4C6M during initial ageing followed by samples with higher fraction of bio-nucleation agent (CNF) during middle-later ageing stage [28]. However, the misalignment of polymer chains occurred in 0C10M-2C8M has caused significantly adverse effect on impact strength measured from 18-24 hours ageing.

The most striking effects of thermal exposure on notched samples can be found at 24 hours ageing stage, mainly related to chain scissioning by thermal degradation which could cause reduction molecular weight or shortening of polymer chains and hence poor resistance to impact loading. This is due to thermally-degraded matrices represented more fracture initiation or crack onset point have been created to facilitate the impact fracture by much lower energy absorption. Lastly, the reduced impact resistance at 72-hours ageing stage majorly related to surface effect based on increased thickness of oxidized layer. The oxidized film played an important role in the reduction of cracking initiation energy since the oxidized matrix becomes stiffer and more brittle than the safe matrix and a stress field was appearing due to the volume variation (resin shrinkage) in the oxidized layer [48]. In other words, the 72-hours-aged specimen was where the microcracks are first initiated in oxidized surface layer with low fracture and damage threshold. Immediately afterwards, the cracking or propagation could be coalesced rapidly from surface-to-core till suddenly cross the ply/laminate, and hence saving the cracking initiation energy contributing to totally lower impact strength throughout this stacking [48].

CONCLUSION

Isothermal ageing has been applied on starch-based polymer and nanocomposites using 60°C, it has prominently enhanced the mechanical properties and water resistance but lowered the impact resistance. The pure TPS has poor heat ageing resistance since polysaccharides is low-molecular-weight carbohydrates with low thermal stability unless the addition of nanofiller (MMT, CNF) to improve the heat ageing resistance of starch films. During the initial ageing stage, the formation of crosslinking (C-C or C-O-C) is beneficial to the improvement of strength and stiffness of samples but caused embrittlement (lower strain and impact strength). Whereas such improvement would be inhibited by silicate layer of MMT based on its plate-like structure with large surface area to block the connection of C•. However, MMT become increasingly qualified to serve as post-cure agent during middle-later ageing stage mainly due to several alteration of arrangement of polymer chains by recrystallization and misalignment through layer effect of MMT. The thermal-ageing favour the increment of the polymer matrix cohesiveness by inducing well-ordered arrangement of moving polymer chains during initial ageing by MMT and during middle-later ageing stage by good nucleation ability of CNF, towards improving modulus and water permeability by increased crystallinity. Unfavorably, the properties of samples would be deteriorated by severely prolonged ageing time at 24-hours mainly caused by reduced crystallinity of samples with higher MMT fraction through misalignment behavior and by decreased molecular weight/crosslinking density on samples with higher CNF fraction through chain-scissioning or thermolysis. However in the 48 hours ageing time, the properties could be further improved by recovery of good recrystallization and molecular rearrangement of remainders respectively. It is suggested avoiding the samples being subjected to thermal ageing till 72 hours considering the oxidized layer with increased thickness would deteriorate the tensile strength and water/impact resistance by its loosening structure, and the disturbance of MMT defects on the recrystallization process of 6C4M-8C2M also should be taken into account. The evolution of the surface morphology behavior from 0/6/18/24/72 hours ageing is considerably well-correlated to the tensile properties. Finally, the brilliance of this research is the nanosize-cellulose fibers showed a potential as post-cure-, nucleating- and antioxidant-agent in starch compounds under thermal ageing in addition to inherently serve as white reinforcing agent in TPS.

ACKNOWLEDGMENT

The authors are grateful for the financial support from Universiti Tunku Abdul Rahman Research Fund 6200/T54.

References

[1] Tee TT, Sin LT, Gobinath R, Bee ST, Hui D, Rahmat AR, Kong I, Fang Q. Investigation of nano-size montomorillnite on enhacing polyvinyl alcohol-starch blends prepared via solution cast approach. *Comp Part B Eng* 2013;47, 238-47.

[2] Le AT, Gacoin A, Li A, Mai TH, Wakil NE. Influence of various starch/hemp mixtures on mechanical and acoustical behavious of starch-hemp composite materials. *Comp Part B Eng* 2015;75, 201-11.

[3] Rahmat AR, Rahman WAWA, Sin LT, Yussuf AA. Approaches to impove compatibility of starch filled polymer system: A review. *Mat Sci Eng C* 2009;29, 2370-7.

[4] Savadekar NR, Mhaske ST. Synthesis of nano cellulose fibers and effect on thermoplastics starch based films. *Carbohyd Polym* 2012;89,146– 51.

[5] Zhang W, Yang XL, Li CY, Liang M, Lua CH, Deng YL. Mechanochemical activation of cellulose and its thermoplastic polyvinyl alcohol ecocomposites with enhanced physicochemical properties. *Carbohyd Polym* 2011;83:257-63.

[6] Ng HM, Sin LT, Tee TT, Bee ST, Hui D, Low CY, Rahmat AR. Extraction of cellulose nanocrytals from plant sources for application as reinforcing agent in polymers. *Comp Part B Eng* 2015;75, 176-200.

[7] Schlemmer D, Angélica RS, Sales MJA. Morphological and thermomechanical characterization of thermoplastic starch/ montmorillonite nanocomposites. *Compos Struct* 2010;92, 2066–70.

[8] Müller P, Renner K, Moczo J, Fekete E, Pukanszky B. Thermoplastic starch/wood composites: Interfacial interactions and functional properties. *Carbohyd Polym* 2014;102, 821–9.

[9] Xie FW, Pollet E, Halley PJ, Averous L. Starch-based nano-biocomposites. *Prog Polym Sci* 2013;38, 1590–628.

[10] Abdul Khalil, HPS., Bhat, AH., & Yusra, AFI. (2012). Green composites from sustainable cellulose nanofibrils: A review. *Carbohydrate Polymers*, 87, 963-979.

[11] Rattanasom N, Prasertsri S. Relationship among mechanical properties, heat ageing resistance, cut growth behaviour and morphology in natural rubber: Partial replacement of clay with various types of carbon black at similar hardness level. *Polym Test* 2009; 28,270–276.

[12] Slavutsky AM, Bertuzzi MA. A phenomenological and thermodynamic study of the water permeation process in corn starch/MMT films. *Carbohyd Polym* 2012; 90, 551– 557.

[13] Guo J, Li XY, Mu CD, Zhang HG, Qin P, Li DF. Freezing-thawing effects on the properties of dialdehyde carboxymethyl cellulose crosslinked gelatin-MMT composite films. *Food Hydrocoll* 2013;33, 273-9.

[14] Vercelheze AES, Fakhouri FM, Dall'Antonia LH, Urbanod A, Youssef EY, Yamashitab F, et al. Properties of baked foams based on cassava starch, sugarcane bagasse fibers and montmorillonite. *Carbohyd Polym* 2012;87, 1302-10.

[15] Santos RMDos, Flauzino-Neto WP, Silvério HA, Martins DF, Dantas NO, Pasquini D. Cellulose nanocrystals from pineapple leaf, a new approach for the reuse of this agro-waste. *Ind Crops Prod* 2013;50,707-14.

[16] Ghanbarzadeh B, Almasi H, Oleyaei SA. A novel modified starch/carboxymethyl/cellulose/montmorillonite bionanocomposite film: structural and physical properties. *Int J Food Eng* 2014;10(1),121-30.

[17] Hietala M, Mathew AP, Oksman K. Bionanocomposites of thermoplastic starch and cellulose nanofibers manufactured using twin-screw extrusion. *Eur Polym J* 2013;49, 950-6.

[18] Ning W, Xingxiang Z, Na H, Shihe Bai. Effect of citric acid and processing on the performance of thermoplastic starch/ montmorillonite nanocomposites. *Carbohyd Polym* 2009;76, 68–73.

[19] Le′veˆque D, Schieffer A, Mavel A, Maire JF. Analysis of how thermal aging affects the long-term mechanical behavior and strength of polymer-matrix composites. *Compos Sci Technol* 2005;65, 395-401.

[20] Pei YM, Wang K, Zhan MS, Xu W, Ding XJ. Thermal-oxidative aging of DGEBA/EPN/LMPA epoxy system: Chemical structure and thermal-mechanical properties. *Polym Degrad Stab* 2011; 96, 1179-86.

[21] Affolter S. *Long-Term Behaviour of Thermoplastic Materials*. NTB, CH-9471 Buchs, Switzerland.

[22] Tamboli SM, Mhaske ST, Kale DD. Crosslinked polyethylene. *Indian J Chem Technol* 2004;11, 853-64.

[23] Martínez-Sanz A, Lopez-Rubio A, Lagaron J. Optimization of the nanofabrication by acid hydrolysis of bacterial cellulose nanowhiskers. *Carbohyd Polym* 2011;85:228-36.

[24] Silverio HA, Neto WPF, Dantas NO, Pasquini D. Extraction and characterization of cellulose nanocrystals from corncob for application as reinforcing agent in nanocomposites. *Ind Crops Prod* 2013;44,427– 36.

[25] Jonoobi M, Aitomäki Y, Mathew AP, Oksman K. Thermoplastic polymer impregnation of cellulose nanofibre networks: Morphology, mechanical and optical properties. *Compos Part A* 2014;58, 30-5.

[26] Xu, SH., Girouard, N., Schueneman, G., Shofner, ML. and Meredith, J. C. (2013). Mechanical and thermal properties of waterborne epoxy composites containing cellulose nanocrystals. *Polymer*. 1-10.

[27] Sehaqui H, Allais M, Zhou Q, Berglund LA. Wood cellulose biocomposites with fibrous structures at micro- and nanoscale. *Compos Sci Technol* 2011;71, 382–7.

[28] Majdzadeh-Ardakani K, Nazari B. Improving the mechanical properties of thermoplastic starch/poly(vinyl alcohol)/clay nanocomposites. *Compos Sci Technol* 2010;70, 1557–63.

[29] Huang MF, Yu JG, Ma XF. High mechanical performance MMT-urea and formamide-plasticized thermoplastic cornstarch biodegradable nanocomposites. *Carbohyd Polym* 2006;63, 393–9.

[30] Saladino ML, Motaung TE, Luyt AS, Spinella A, Nasillo G, Caponetti E. The effect of silica nanoparticles on the morphology, mechanical properties and thermal degradation kinetics of PMMA. *Polym Degrad Stab* 2012;97, 452-9.

[31] Majeed K, Jawaid M, Hassan A, Abu Bakar A, Abdul Khalil HPS, Salema AA, et al. Potential materials for food packaging from nanoclay/natural fibres filled hybrid composites. *Mater Des* 2013;46, 391–410.

[32] Espigulé E, Puigvert X, Vilaseca F, Mendez JA, Mutjé P, Girones J. Thermoplastic starch-based composites reinforced with rape fibers: water uptake and thermomechanical properties. *Bioresourc* 2013;8(2), 2620-2630.

[33] Jain A, Vijayan K. Thermal Aging of Twaron Fibers. *High Perform Polym* 2003;15,105-29.

[34] Boubakri A, Haddar N, Elleuch K, Bienvenu Y. *Influence of thermal ageing on tensile and creep behavior of thermoplastic polyurethane*. C. R. Mecanique 2011;339, 666-73.

[35] Srithep, Y, Ellingham, T, Peng, J, Sabo, R, & Clemons, C. (2013). Melt compounding of poly (3-hydroxybutyrate-co-3-hydroxyvalerate)/nanofibrillated cellulose nanocomposites. *Polymer Degradation and Stability*, 98, 1439-1449.

[36] Espino-Pérez E, Bras J, Ducruet V, Guinault A, Dufresne A, Domenek S. Influence of chemical surface modification of cellulose nanowhiskers on thermal, mechanical, and barrier properties of poly(lactide) based bionanocomposites. *Europ Polym J* 2013;49, 3144–54.

[37] Seguchi T, Tamura K, Shimada A, Sugimoto M, Kudoh H. Mechanism of antioxidant interaction on polymer oxidation by thermal and radiation ageing. *Radiat Phys and Chem* 2012;81, 1747–51.

[38] Abdollahi M, Alboofetileh M, Rezaei M, Behrooz R. Comparing physico-mechanical and thermal properties of alginate nanocomposite films reinforced with organic and/or inorganic nanofillers. *Food Hydroc* 2013; 32, 416-24.

[39] Bee ST, Hassan A, Ratnam CT, Tee TT, Lee TS. Investigation of nano-size montmorillonite on electron beam irradiated flame retardant polyethylene and ethylene vinyl acetate blends. *Nucl Instr Meth Phys Res B* 2013; 299, 42–50.

[40] Huang MF, Yu JG, Ma XF. Studics on the properties of Montmorillonite-reinforced thermoplastic starch composites. *Polym* 2004;45, 7017–23.

[41] Chuayjuljit S, Hosililak S, Athisart A. Thermoplastic Cassava Starch/Sorbitol-Modified Montmorillonite Nanocomposites Blended with Low Density Polyethylene: Properties and Biodegradability Study. *Met Mater Miner* 2009;19(1), 59-65.

[42] Heydari A, Alemzadeh I, Vossoughi M. Influence of Glycerol and Clay Contents on Biodegradability of Corn Starch Nanocomposites. *IJE Transact B: Appl* 2014;27(2), 203-14.

[43] Acharya SK, Mishra P, Mehar SK. Effect of surface treatment on the mechanical properties of bagasse fiber reinforced polymer composite. *Bioresource* 2011;6(3), 3155-65.

[44] Shang L, Han GP, Zhu FZ, Ding JS, Shupe T, Wang QW, et al. High-density polyethylene-based composites with pressure-treated wood fibers. *Bioresource* 2012;7(4), 5181-9.

[45] Sarifuddin N, Ismail H. Comparative study on the effect of bentonite or feldspar filled low-density polyethylene/thermoplastic sago starch/kenaf core fiber composites. *Bioresource* 2013;8(3), 4238-57.

[46] Faruk O, Bledzki AK, Fink HP, Sain M. Biocomposites reinforced with natural fibers: 2000-2010. *Prog Polym Sci* 2012;37, 1552–96.

[47] Yang HS, Gardner DJ, Nader JW. Characteristic impact resistance model analysis of cellulose nanofibril-filled polypropylene composites. *Compos Part A* 2011, 42, 2028–2035.

[48] Co´rdoba A, Cue´llar N, Gonza´lez M, Medina J. The plasticizing effect of alginate on the thermoplastic starch/glycerin blends. *Carbohyd Polym* 2008;73,409-16.

In: Polymer-Matrix Composites ISBN: 978-1-53613-398-1
Editors: Elias Rice and Beth Sparks

Chapter 4

AN ANALYSIS OF ELECTRON BEAM IRRADIATION INTERACTION WITH TRANSITION METAL OXIDE AND SALT PRESENCE IN THERMOPLASTIC STARCH

Bee Soo-Tueen[1,*], Lee Tin Sin[1,†], C. T. Ratnam[2], Ezra Joel Tharmapal[1] and A. R. Rahmat[3]

[1]Department of Chemical Engineering,
Lee Kong Chian Faculty of Engineering and Science,
Universiti Tunku Abdul Rahman, Jalan Sungai Long,
Bandar Sungai Long, Cheras, Selangor, Malaysia
[2]Radiation Processing Technology Division,
Malaysian Nuclear Agency, Bangi, Selangor, Malaysia
[3]Department of Polymer Engineering,
Faculty of Chemical Engineering,
Universiti Teknologi Malaysia, Johor, Malaysia

* Corresponding Author Email: direct.beest@gmail.com.
† Corresponding Author Email: direct.tinsin@gmail.com.

ABSTRACT

This research investigated the effects of different loading level ratios of $CuO/MnCl_2$ and electron beam irradiation on the properties of thermoplastic starch (TPS). The increasing of manganese (II) chloride loading level has gradually improved the tensile strength and Young's modulus of all un-irradiated and irradiated $CuO/MnCl_2$ added TPS. This is because the manganese (II) chloride could form a strong interaction with starch molecules in TPS matrix. Besides, the increasing of copper (II) oxide loading level has significantly improved the thermal stability of TPS matrix by delaying the thermal degradation temperatures of TPS matrix. The copper (II) oxide in TPS matrix could enhance the thermal stability of TPS by acting as inorganic retarder during combustion process. The application of electron beam irradiation could interact well with manganese (II) chloride particles in TPS matrix by inducing the crystallinity of TPS matrix. This interaction effect could increase the crystallinity of TPS by converging the random polymer chains arrangement in amorphous region of TPS matrix into more ordered structure.

Keywords: thermoplastic starch, irradiation, mechanical properties, morphology

INTRODUCTION

Recently, many polymer related researches have been focusing on the area of biodegradable polymer. The main reasons of such development are mainly because of depletion of petroleum resources and the inherent non-biodegradability of petroleum-based materials [1]. One of the abundant renewable resources, i.e., starch is considered to be the high potential material to produce biodegradable polymer materials. There are many advantages of starch include its inherent biodegradable characteristic, ubiquitous availability, relatively low cost and the ease for chemical modifications [2-4].

Generally, native starch presents in granular structure and can be modified into thermoplastic starch (TPS) by processing under high temperature and shear conditions in presence of plasticizers [5-6]. Over the decades, the processing technology of TPS has been improved to meet the requirement for industrialization. TPS is able to be processed existing technology used for synthetic polymers like extrusion, blowing, thermo-compression or injection molding [7-9]. Nevertheless, TPS materials exhibit some disadvantages such as their hydrophilic behaviour and weak mechanical and barrier properties compared to synthetic polymers [10]. Many researches have been done to improve such limitation by blending TPS with organic and mineral fillers. These fillers reinforce biopolymeric matrixes and led to develop films with special properties due to the synergic effect between the components [11]. TPS–composite films have a potential use as packaging materials for food, cosmetics, and pharmaceutics products, among others. However, when polymer reinforced with micro-sized fillers, these fillers could act as stress concentrators stress concentrators affecting the mechanical performance of the final materials [12]. A common method to maintain the usage of fillers without the cause of lost of properties is through the application of crosslinking onto the polymer blends. Radiation crosslinking is one of the methods used to improving the properties of polymer by introducing the formation of crosslinking networks in polymer matrix [13-15]. Recently, the importance of irradiation technique has been significantly increased as an alternative method to induce the crosslinking networks in polymers matrix besides application of chemical crosslinking agent. This is due to the application of irradiation technique could eliminate the remaining chemical residues from chemical crosslinking agents in polymer which might be harmful to the consumers. Generally, there are various forms of high energy irradiations such as electron beam irradiation, gamma rays irradiation and X-ray irradiation which are currently used to modify the properties of polymer materials. The application of irradiation technique on polymers can induce the formation of crosslinking by generating polymeric free radicals in polymer matrix. These free radicals could crosslink two long macromolecule chains by forming C-C intermolecular bonds to modify the properties of polymers. In general, two important mechanisms such as crosslinking process and chain scissioning reactions would be occurred during the irradiation process of polymers. The dominant mechanism could affect the changes in matrix structure and also mechanical properties of polymers. The mechanical-structural properties of polymer matrix could be significantly enhanced when the crosslinking reaction is more dominance than chain scissioning reaction during the irradiation process. This is due to the formation of crosslinked networks in polymer matrix could significantly increase the molecular weights of polymers. On the other hand, the chain scissioning reaction is pre-dominated during the irradiation process. The polymer matrix could be severely degraded by breaking down the macromolecules in polymer into smaller sizes. Thus, this could further reduce the molecular weight of polymer and weaken the mechanical-structural properties of polymer [16]. In order to reduce the substantial generated of radicals that would lead to chain scissioning, addition of transition metal complexes can be helpful as radical scavengers. The transition metal such as copper, manganese, iron, zinc and etc. could react with traces of radical resulted from the attacks on the starch molecules [16]. As compared to the other radical scavengers like glycerol, transition metal complex can also provide some reinforcement to the TPS resulted from the rigidity characteristics of the materials. In this study, the three elements of copper oxide, manganese chloride and electron beam irradiation interactions were analyzed to determine their influences on the physicomechanical properties of TPS.

Experimental

Materials

In this study, the regular tapioca starch was obtained from superstone in Malaysia. Tapioca starch was used as the base polymer to produce thermoplastic starch (TPS). Glycerol of BP/USP grade with 99.5% purity was purchased from Sigma Aldrich, Malaysia. Glycerol was used as plasticizer in this research. Copper (II) oxide of AR grade with 99.9% purity and manganese (II) chloride with AR grade of 99% purity were supplied by Sigma Aldrich, Malaysia.

Samples Preparation

The mixtures of tapioca starch, glycerol, copper (II) oxide (CuO) and manganese (II) chloride ($MnCl_2$) were prepared using Brabender mixer. The loading level ratio of tapioca starch, glycerol and fillers (copper (II) oxide and manganese (II) chloride) were maintained at 100 phr: 30 phr: 10 phr. The loading levels of copper (II) oxide and manganese (II) chloride were varied from 0 phr to 10 phr. Initially, copper (II) oxide and manganese (II) chloride were pre-mixed with tapioca starch to achieve a good dispersion of copper (II) oxide and manganese (II) chloride particles in tapioca starch powder. Later, glycerol was added into the pre-mixed tapioca starch-CuO-$MnCl_2$ powder and mixed for 15 minutes to produce thermoplastic starch mixtures. These CuO-$MnCl_2$ added thermoplastic starch mixtures were further processed and melt blended in a Brabender mixer at heating temperature of 110°C for a mixing time of 8 minutes. After that, the blended CuO-$MnCl_2$ added thermoplastic starch samples were compression moulded into 3-mm thickness sheet using hot press machine under heating temperature of 130°C. During the compression moulding process, the blended CuO-$MnCl_2$ added thermoplastic starch samples were initially preheated inside hot press machine at heating temperature of 130°C for 5 minutes. Later, the preheated samples were compression moulded at heating temperature and pressure of 130°C and 10 MPa, respectively for another 5 minutes. The compression moulded samples were further cooled down to room temperature under pressure of 10 MPa with a cooling rate of 15°C/min. The 3-mm thickness sheets were then cut into required shape using sample cutter. The cut samples were then further electron beam irradiated to 20 kGy, 40 kGy and 60 kGy using an electron beam accelerator at 20 kGy per pass.

Tensile Test

The tensile test was performed to investigate the tensile properties of samples under stretching. The 3-mm thickness sheets were cut into dumbbell shape using tensile dumbbell cutter in accordance to the ASTM D1822. The gauge thickness of samples was measured prior the tensile testing. The tensile test was conducted by using Instron Universal Testing Machine (model 4302 Series IX) with crosshead speed of 50 mm/min. The dumbbell specimens were tested at room temperature of 25 °C under load of 2 kN. The results of tensile strength, elongation at break and Young's modulus were taken as an average of five specimens.

Infrared Spectroscopy Analysis

Fourier Transform Infrared- Attenuated total reflection analysis was conducted using the Thermo-Fisher Nicolet iS10. The samples were ground into powder form and placed onto sample holder for scanning in the region of 4000–400 cm^{-1}.

Scanning Electron Microscopy (SEM) Analysis

Scanning Electron Microscopy (SEM) analysis was carried out using Hitachi S-3400N Scanning Electronic Microscopy (SEM). The fractured surfaces of the samples were initially cut into small portions before performing the SEM scanning. The cut samples were then placed and mounted onto the copper stub with the fractured surface is facing upwards. After that, the samples were coated with a thin layer of palladium and gold using EMITECH SC7620 Sputter Coater in order to allow the samples to be more conductive towards electrons. The samples scanning were carried out under the magnification of 1000 times, 3000 times, 5000 times at an electron beam voltage of 15 kV.

Thermoravimetric Analysis (TGA)

Thermogravimetric analysis (TGA) test was carried out using Perkin Elmer Thermogravimetric Analyzer Pyris 1 TGA. Before the TGA testing, nitrogen gas was purged into the combustion column with flow rate of 25 ml/min to remove the air inside combustion column. The thermoplastic starch samples were cut into smaller pieces with the weight range of 1 mg to 2 mg. The cut samples with weight approximately 1 mg to 2 mg were heated from room temperature to 850°C using a platinum crucible in a micro-furnace at a heating rate of 20°C/min under nitrogen atmosphere. The decomposition temperature at 5%, 15%, 25%, 35%, 45% and 55% weight loss as well as the remaining of char residue of the samples were recorded and further analysed.

X-Ray Diffraction (XRD) Test

The crystalline structure of thermoplastic starch samples was examinated by using X-ray Diffraction Shimadzu XRD 6000 diffractometer with Cu-Kα radiation ($\lambda = 1.542$ Å) for a 2θ range of 0° to 40° at a scanning rate of 1.2°/min. The Cu-Kα radiation generator was operated at acceleration voltage and current of 40 kV and 30 mA. The degree of crystallinity was calculated as the percentage of the scattered intensity of the crystalline phase over the scattered intensity of the crystalline and amorphous phase. The crystalline size, L was derived from the peak angle (2θ) in the XRD diffractogram according to the Scherrer equation (L).

$$L = \frac{k\lambda}{\beta \cos\theta} \tag{1}$$

where k is Scherrer constant (usually taken as 0.9), β is the full width at half maxima (FWHM) of the diffraction peak, λ is 1.542 Å and θ is the

Bragg angle in radians. The d-spacing of crystallite structure were calculated using the Bragg's equation as shown in equation (2). On the other hand, the inter-chain separation, R of crystalline structure was calculated using Klug and Alexander equation as shown in equation (3).

$$d = \frac{\lambda}{2\sin\theta} \tag{2}$$

$$R = \frac{5\lambda}{8\sin\theta} \tag{3}$$

Results and Discussion

Mechanical Properties Analysis

Tensile Strength

Figure 1(a) illustrates the effect of electron beam irradiation on the tensile strength of thermoplastic starch (TPS) added with various loading level ratios of copper (II) oxide/ manganese (II) chloride ($CuO/MnCl_2$). By referring to Figure 1(a), all the un-irradiated and irradiated 10 phr manganese (II) chloride added TPS samples exhibit the highest tensile strength in comparing to the all $CuO/MnCl_2$ added TPS samples. Whereas, all the un-irradiated and irradiated 10 phr copper (II) oxide added TPS samples show the poorest tensile strength. The increasing of manganese (II) chloride loading level in replacing the copper (II) oxide have gradually increased the tensile strength of un-irradiated TPS samples. This also indicates that the addition of manganese (II) chloride particles could provide better reinforcement effect than copper (II) oxide particles to thermoplastic starch matrix. This is due to the presence of manganese (II) chloride in thermoplastic starch matrix could form strong interaction bonding with starch molecules [17]. This could enhance the interaction bonding between manganese (II) chloride particles and thermoplastic starch matrix by replacing the intramolecular and intermolecular hydrogen bonding of starch matrix [17]. Thus, this also could further increase the resistance ability of thermoplastic starch matrix when subjected to straining stress. By referring to FTIR results in Figure 2 and Table 1, the deformation vibration of hydroxyl group (O-H) of starch molecules in 10 phr copper (II) oxide added TPS was found to occur at 3277.62 cm^{-1} as shown in Figure 2 and Table 1. The absorption frequency of hydroxyl group of un-irradiated TPS samples was found to be gradually shifted to lower value from 3277.62 cm^{-1} to 3259.19 cm^{-1} with increasing of manganese (II) chloride loading level from 0 phr to 10 phr as shown in Table 1. This observation also indicates improvement in strength of the O-H bonding of starch molecules after the addition of manganese (II) chloride particles. The reduction of absorption frequency has been reported by Tee et al. [18] which is known as "red shift" effect. Such effect is contributed by the addition of manganese (II) chloride forms a stronger interaction with starch molecules that enhancing the O-H bonding of starch molecules. Thus, the strong interaction between manganese (II) chloride particles and starch molecules could significantly enhance the tensile strength of thermoplastic starch matrix [17-18].

By referring to Figure 1(a), the application of electron beam irradiation on all the CuO/ $MnCl_2$ added TPS samples were observed to gradually decrease the tensile strength. This is due to the application of electron beam irradiation could attack the hydroxyl group (O-H) and C-O bonding of starch molecules in TPS to form ether or ester intermolecular linkages between hydroxyl groups on starch molecules [19]. The formation of these ether and ester linkages between hydroxyl groups in thermoplastic starch matrix by electron beam irradiation could reduce the intramolecular and intermolecular bonding such as hydrogen bonding between the starch molecules [17]. This decrement in intramolecular and intermolecular bonding in thermoplastic starch matrix could significantly reduce the ability of thermoplastic starch matrix in resisting the stress applied on the

TPS samples. Thus, the tensile strength of $CuO/MnCl_2$ added TPS samples have been significantly reduced by the application of electron beam irradiation. This also can be further proved by increasing the absorption frequency of hydroxyl groups in all $CuO/MnCl_2$ added TPS matrix after the application of electron beam irradiation as shown in Table 1. The weakening effect of intra- and intermolecular bonding between the starch molecules in thermoplastic starch matrix could significantly cause the hydroxyl groups of starch to be stronger. Besides, the increasing of electron beam irradiation dosages on TPS samples also found to gradually reduce the tensile strength of all CuO/ $MnCl_2$ added TPS samples. This is due to the increasing of electron beam irradiation dosages could induce the formation of intermolecular linkages between hydroxyl groups inside thermoplastic starch matrix where this could further reduce the intramolecular and intermolecular bonding inside TPS matrix. The reduction in intramolecular and intermolecular bonding in thermoplastic starch matrix could decrease the resistance ability of polymer matrix towards the applied straining stress [20]. Thus, this could further reduce the tensile strength of all TPS samples.

Table 1. Absorption frequency (wavenumber) of hydroxyl group (O-H) of all copper (II) oxide/ manganese (II) chloride added thermoplastic starch samples

Sample formulation	Wavenumber, cm^{-1}			
	0 kGy	20 kGy	40 kGy	60 kGy
10 phr CuO: 0 phr $MnCl_2$	3277.62	3279.83	3280.86	3291.11
8 phr CuO: 2 phr $MnCl_2$	3262.66	3261.92	3261.73	3260.81
5 phr CuO: 5 phr $MnCl_2$	3261.11	3261.75	3260.07	3262.02
2 phr CuO: 8 phr $MnCl_2$	3260.31	3262.23	3261.37	3262.22
0 phr CuO: 10 phr $MnCl_2$	3259.19	3261.70	3261.92	3260.32

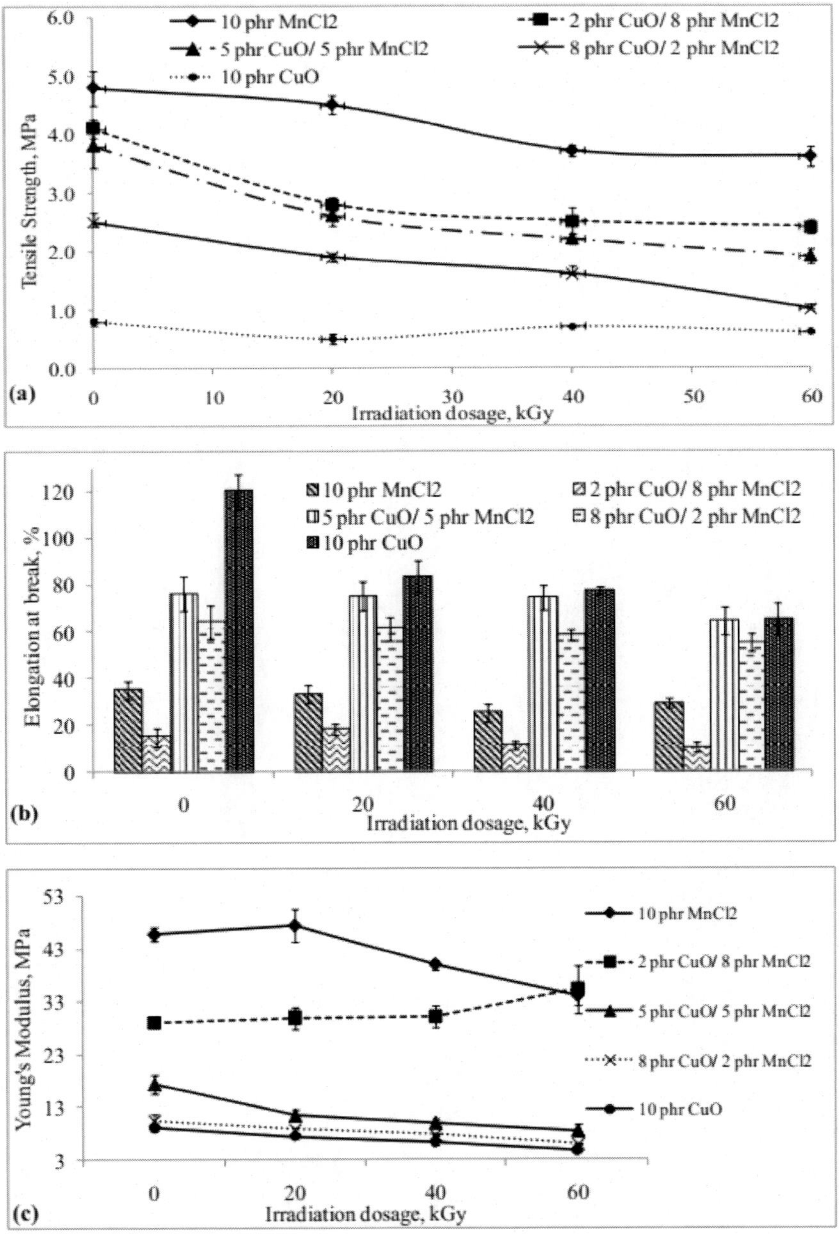

Figure 1. (a) Tensile strength, (b) elongation at break and (c) Young's modulus of thermoplastic starch added with various loading levels of Manganese (II) chloride ($MnCl_2$)/copper (II) oxide (CuO) under various electron beam irradiation dosages.

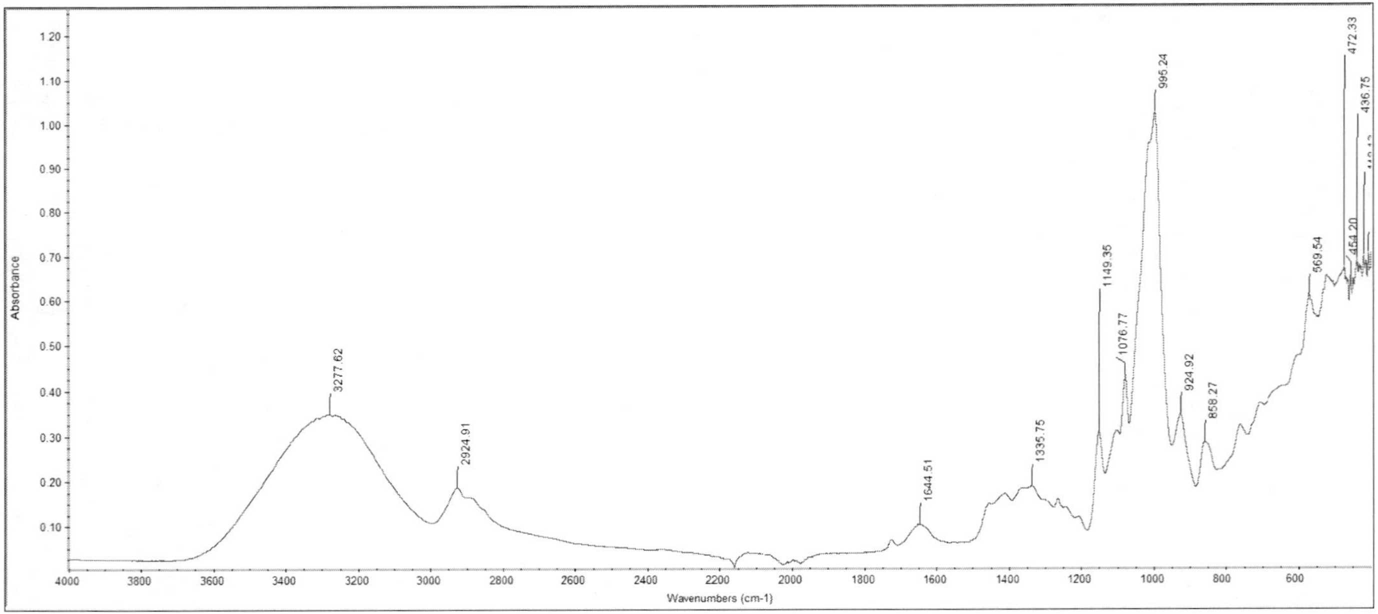

Figure 2. The FT-IR spectra of 10 phr copper (II) oxide added thermoplastic starch.

Elongation at Break

By referring to Figure 1(b), the elongation at break of un-irradiated and all irradiated thermoplastic starch (TPS) samples was observed to gradually increase with increasing of copper (II) oxide loading level from 0 phr to 10 phr in replacing the manganese (II) chloride. In other words, the decreasing of manganese (II) chloride could significantly increase the matrix deformation ability of thermoplastic starch by increasing the elongation at break of TPS. This might be due to addition of manganese (II) chloride particles into thermoplastic starch matrix could form strong interaction effect with starch molecules [17]. The occurrence of strong interaction between manganese (II) chloride and starch molecules could decrease the deformation ability of thermoplastic starch matrix when subjected to straining stress. This caused the elongation at break of TPS to gradually increase with decreasing the loading level of manganese (II) chloride in thermoplastic starch matrix (or with increasing of copper (II) oxide loading level). Besides, the addition of copper (II) oxide particles into thermoplastic starch matrix could induce the formation of a very small degree of crosslinking in polymer matrix during samples compounding [21]. The very low degree of crosslinking formed in thermoplastic starch matrix by copper (II) oxide could help in extending the elongation ability of TPS during straining and thus increase the elongation ability of thermoplastic starch.

According to Figure 1(b), the increasing of electron beam irradiation dosages from 0 kGy to 60 kGy has gradually reduced the elongation at break of all $CuO/MnCl_2$ added TPS samples. This is due to the application of electron beam irradiation on thermoplastic starch matrix could induce the formation of crosslinking in thermoplastic starch matrix by generating free radicals in TPS matrix [22-24]. The formation of crosslinking inside TPS matrix could restrict the mobility or slippage effect of starch molecules when subjected to straining stress [25]. The intermolecular linkages between hydroxyl groups on starch molecules (also known as crosslinking networks) could reduce the deformation ability of TPS matrix during straining and thus this could further reduce the elongation at break.

By referring to Figure 1(b), the effect of electron beam irradiation on elongation at break of TPS added 10 phr manganese (II) chloride and $CuO/MnCl_2$ was observed to be small in comparing to 10 phr copper (II) oxide added TPS. This is due to the copper (II) oxide in 10 phr copper (II) oxide added TPS would receive two electrons from electron beam accelerator and form Cu and ion O^{2-} [26]. These ions O^{2-} would attack the starch molecules to form free radicals by releasing water vapour. The free radicals formed would further react with each other by forming a crosslinking network between two starch molecules [26]. The degree of crosslinking networks formed in the polymer matrix of 10 phr copper (II) oxide added TPS by electron beam irradiation was significantly higher than the degree of crosslinking in 10 phr manganese (II) chloride and all $CuO/MnCl_2$ added TPS samples. Thus, the decrement in elongation at break of 10 phr copper (II) oxide added TPS samples was found to be more significantly than the other TPS samples.

Young's Modulus

By referring to Figure 1(c), the Young's modulus of all un-irradiated and irradiated $CuO/MnCl_2$ added thermoplastic starch (TPS) samples was observed to gradually increase with increasing of manganese (II) chloride loading level from 0 phr to 10 phr. This also indicates the addition of manganese (II) chloride could highly increase the Young's modulus (stiffness) of the TPS samples. This could be due to the manganese (II) chloride in TPS matrix could form a strong interaction effect with starch molecules [17]. The strong interaction formed between the manganese (II) chloride particles and starch particles could highly enhance the interfacial adhesion between manganese (II) chloride particles and TPS matrix. This could lead to better dispersion of manganese (II) chloride particles in TPS matrix and thus further improve the stiffness of the TPS samples. On the other hand, the addition of copper (II) oxide particles into TPS matrix did not exhibit a good interaction effect between the copper (II) oxide particles and starch molecules. The poor interaction between the copper (II) oxide particles and TPS matrix has caused the copper (II) oxide particles to severely agglomerate into larger aggregates [20]. The occurrence of copper (II) oxide aggregates (agglomerated) on fractured surface of 10 phr copper (II) oxide added TPS can be clearly observed by SEM test as shown in Figures 3(a) and 3(b). The poor interfacial adhesion effect between the copper (II) oxide aggregates and thermoplastic starch matrix is mainly due to the irregularity shapes of the copper (II) oxide aggregates. This could lead to the occurrence of cavities in between the interfaces of copper (II) oxide aggregates and TPS matrix and cause the stress applied on the TPS matrix unable to effectively and evenly transfer throughout the whole TPS matrix [25]. Thus, the Young's modulus was significantly reduced with increasing of copper (II) oxide loading level. By comparing the Figure 3(a) and 3(c), the manganese (II) chloride particles were observed to be well disperse in 10 phr manganese (II) chloride added TPS matrix without the presence of manganese (II) chloride particles agglomeration. Thus, the Young's modulus of 10 phr manganese (II) chloride added TPS was found to be the highest due to the good dispersion and interaction of manganese (II) chloride particles with TPS matrix.

By referring to Figure 1(c), the increasing of electron beam irradiation dosages up to 60 kGy was found to gradually decrease the Young's modulus of all $CuO/MnCl_2$ added TPS samples. The application of electron beam irradiation could induce the degree of crosslinking formed in TPS by releasing the water vapor. The water vapor released during the electron beam irradiation process would trap inside the TPS matrix and form bubbles or voids as illustrated in Figure 3(e). The occurrence of bubbles or voids in TPS matrix could act as stress concentration point (weak point) when subjected to straining stress and cause the stress applied on the thermoplastic starch matrix unable to evenly transfer throughout the TPS matrix. Thus, the Young's modulus (stiffness) of all the TPS was decreased with increasing of electron beam irradiation dosages.

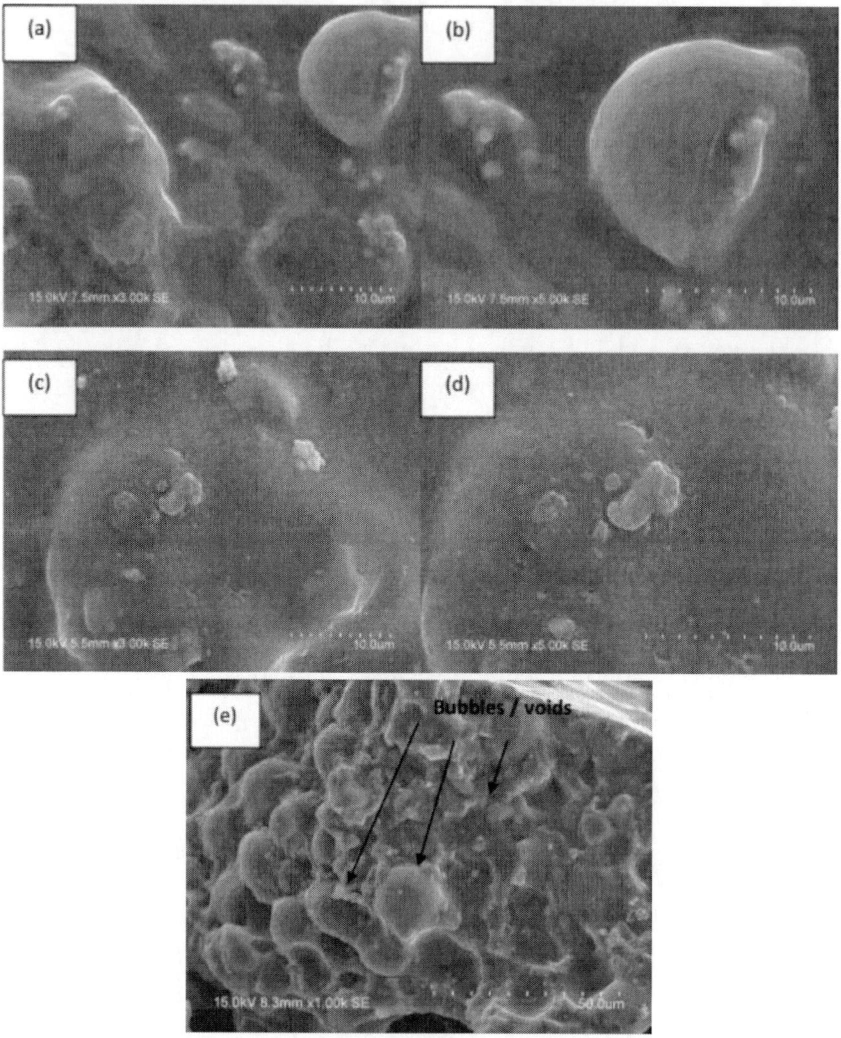

Figure 3. SEM micrographs of surface morphologies for (a) and (b) un-irradiated 10 phr copper (II) oxide added thermoplastic starch under the magnification of 3000 times and 5000 times, respectively, (c) and (d) un-irradiated 10 phr manganese (II) chloride added thermoplastic starch under the magnification of 3000 times and 5000 times, respectively and (e) 8 phr copper (II) oxide + 2 phr manganese (II) chloride added thermoplastic starch when irradiated up to 60 kGy.

Thermal Properties

The thermal stabilities of all CuO/$MnCl_2$ added TPS samples when subjected to irradiation dosages of 20 kGy and 60 kGy were studies in nitrogen environment. Figure 4 illustrates the TGA curves of TPS added with different loading level ratios of CuO/$MnCl_2$ when subjected to irradiation dosages of 20 kGy and 60 kGy. By referring to Figure 4, two thermal degradation stages were occurred during the thermal degradation process of all CuO/$MnCl_2$ added TPS samples. The first degradation stage is observed to occur at the temperature range of 80°C to 200°C and the existence of this degradation stage is due to loss of moistures (water evaporation) and glycerol content (glycerol evaporation) inside thermoplastic starch samples during thermal decomposition process [28-30]. Besides, the dehydration of $MnCl_2$ particles inside TPS samples also could contribute to the weight loss of first stage degradation. Furthermore, the weight loss during first degradation stage of all 10 phr $MnCl_2$ added TPS samples is found to be higher than other TPS samples (5 phr $MnCl_2$ + 5 phr CuO and 10 phr CuO added TPS). This is due to the higher content of manganese (II) chloride inside 10 phr $MnCl_2$ added thermoplastic starch samples could absorb water vapor from surrounding during sample compounding and sample storage. The dehydration of $MnCl_2$ particles in TPS matrix have highly contributed to the highest weight loss at first stage degradation of 10 phr $MnCl_2$ added TPS samples.

By referring to Figure 4, the second degradation stage is observed to be occurred at temperature range of 255°C to 480°C and this degradation stage is attributed to the thermal decomposition of starch in thermoplastic starch matrix. This results observation is observed to be consistent with the finding found by Kumar et al. [29] and Kaewtatip & Thongmee [30]. However, the weight loss at second degradation stage of 10 phr CuO added TPS samples is observed to be the highest in comparing to other TPS samples. This also indicates that the occurrence of $MnCl_2$ particles in TPS matrix could reduce the weight loss at second degradation stage. In other words, the addition of copper (II) oxide particles into thermoplastic starch matrix could provide more effective degradation action than manganese

(II) chloride in degrading TPS matrix under heating by copper-catalyzing the thermoplastic starch matrix [26].

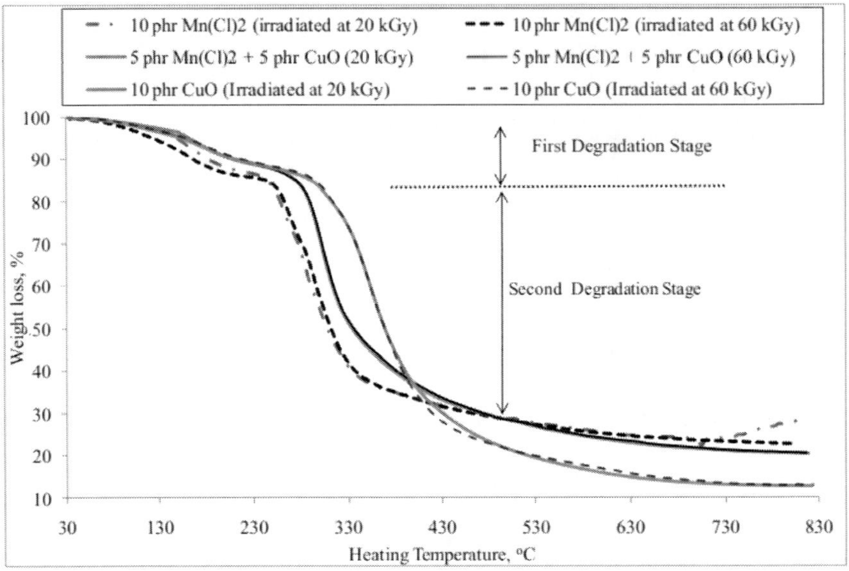

Figure 4. TGA curves of thermoplastic starch added with various loading level ratio of copper (II) oxide/manganese (II) chloride when subjected irradiation dosages of 20 kGy and 60 kGy, respectively.

The change in thermal decomposition behaviour of thermoplastic starch samples also can be studied by referring to thermal degradation temperatures at 5%, 15% 25%, 35%, 45% and 55% weight losses as tabulated in Table 2. The increasing of copper (II) oxide loading level from 0 phr to 5 phr in replacing manganese (II) chloride loading level was observed to gradually increase the all the thermal degradation temperatures ($T_{5\%}$, $T_{15\%}$, $T_{25\%}$, $T_{35\%}$, $T_{45\%}$ and $T_{55\%}$) of TPS samples. This indicates that the increasing of copper (II) oxide loading level could gradually increase the thermal stability of TPS by delaying the thermal degradation temperatures. This is due to the addition of copper (II) oxide into polymer matrix could significantly delay the thermal degradation to higher thermal degradation temperature by acting as inorganic catalyst during combustion process [21, 31]. As shown in Table 2, with further increasing of copper

(II) oxide loading level in replacing the manganese (II) chloride content inside thermoplastic starch matrix (from 5 phr to 10 phr) have significantly shifted the $T_{15\%}$, $T_{25\%}$, $T_{35\%}$, $T_{45\%}$ and $T_{55\%}$ of thermoplastic starch samples to higher temperatures. It was revealed that the higher loading level of copper (II) oxide particles in TPS matrix could further enhance the thermal stability of TPS samples by increasing the thermal decomposition temperature of thermoplastic starch samples.

Table 2. The remaining char residues and thermal degradation temperature of thermoplastic starch added with copper (II) oxide, copper (II) oxide/manganese (II) chloride and manganese (II) chloride at various weight loss under irradiation dosages of 20 kGy and 60 kGy

Loading level of Manganese (II) chloride: copper (II) oxide (phr: phr)	Irradiation dosages, kGy	Thermal Degradation Temperature at various weight loss, °C						Remaining char residue, %
		$T_{5\%}$	$T_{15\%}$	$T_{25\%}$	$T_{35\%}$	$T_{45\%}$	$T_{55\%}$	
10: 0	20	148.8 ± 1.2	244.3 ± 0.6	267.1 ± 0.9	283.9 ± 0.8	298.8 ± 0.8	318.7 ± 0.6	27.3 ± 0.8
10: 0	60	122.8 ± 0.8	242.3 ± 0.8	270.9 ± 1.1	289.4 ± 1.2	303.6 ± 0.9	321.4 ± 0.9	22.7 ± 1.2
5: 5	20	161.0 ± 1.1	273.6 ± 0.7	295.2 ± 0.8	306.3 ± 1.1	320.9 ± 0.8	351.6 ± 0.8	20.3 ± 1.1
5: 5	60	154.5 ± 1.0	273.0 ± 0.8	296.3 ± 0.9	308.0 ± 0.9	322.6 ± 1.2	355.7 ± 1.1	20.6 ± 0.9
0: 10	20	153.5 ± 1.0	289.7 ± 0.9	327.4 ± 1.0	345.9 ± 0.9	360.3 ± 0.5	378.0 ± 0.9	12.6 ± 0.5
0: 10	60	158.0 ± 0.8	294.2 ± 0.9	328.1 ± 1.1	346.0 ± 1.0	360.6 ± 0.6	377.3 ± 0.8	12.9 ± 0.6

Besides, the effect of increasing the electron beam irradiation on thermal degradation temperature of thermoplastic starch also can be observed in Table 2. The increasing of electron beam irradiation dosages from 20 kGy to 60 kGy has slightly decreased the $T_{5\%}$ and $T_{15\%}$ of all $MnCl_2$ added TPS samples (10 phr $MnCl_2$ and 5 phr $MnCl_2$ + 5 phr CuO added TPS samples). However, the $T_{5\%}$ and $T_{15\%}$ of 10 phr CuO added TPS was slightly increased as the irradiation dosages increased from 20 kGy to 60 kGy. This is due to the increasing of electron beam irradiation dosage up to 60 kGy on could induce the water absorption of manganese (II) chloride added TPS samples during the irradiation process. The more water absorbed into TPS matrix could significantly increase the weight loss at first stage of degradation (due to water evaporation) at lower thermal degradation temperature. In other words, the application of high irradiation dosage (up to 60 kGy) could reduce the $T_{5\%}$ and $T_{15\%}$ of $MnCl_2$ added thermoplastic starch samples due to loss of high moisture content. However, the increasing of electron beam irradiation from 20 kGy to 60 kGy showed an opposite result on $T_{5\%}$ and $T_{15\%}$ of 10 phr copper (II) oxide added TPS. This is due to the copper (II) oxide added TPS did not further promote water absorption process during electron beam irradiation process. The lower moisture content of 10 phr copper (II) oxide added TPS could delay the $T_{5\%}$ and $T_{15\%}$ to higher temperatures.

According to Table 2, the $T_{25\%}$, $T_{35\%}$, $T_{45\%}$ and $T_{55\%}$ of all thermoplastic starch samples were observed to gradually increase when the irradiation dosages increased from 20 kGy to 60 kGy. This is due to the application of higher irradiation dosage (up to 60 kGy) on TPS matrix tended to cause more cross-link bonds at random locations in the tapioca starch granules. The occurrence of cross-link bonds in thermoplastic starch matrix is mainly attributed to the formation of ester inter-molecular linkages between the hydroxyl groups inside TPS matrix [19]. The higher degree of cross-link bonds in TPS samples due to higher irradiation dosages could exhibit better thermal stability than 20 kGy irradiated TPS samples by delaying the thermal degradation temperatures. However, the increasing of irradiation dosages from 20 kGy to 60 kGy was observed to pose a less significant effect in increasing the thermal degradation temperatures at weight losses of 35%, 45% and 55% of 10 phr copper (II) oxide added TPS. This might be due to the copper (II) oxide in TPS matrix could slightly accelerate degradation of thermoplastic under high temperature environment [32]. At high temperature, the ion Cu^{2+} (from copper (II) oxide) in thermoplastic starch matrix could decompose the –COOH groups by forming –CO and –COO free radicals where these reactions could help in accelerating the degradation of starch molecules.

In addition, the increasing of copper (II) oxide particles inside TPS matrix also found to gradually decrease the remaining char residues formed after the combustion process as shown in Table 2. The lowering in remaining char residues of higher copper (II) oxide content added TPS samples after the thermal decomposition process in nitrogen atmosphere are believed to be responsible by the catalytic effect of copper (II) oxide on the degradation reaction of TPS matrix as discussed earlier [32]. The catalytic effect of copper (II) oxide in TPS matrix did not provide significant effect in promoting the formation of char residues. However, the increasing of manganese (II) chloride loading level from 0 phr to 10 phr in TPS matrix has gradually increase the remaining char residue after the thermal decomposition process as indicated in Table 2. The addition of transition metal chloride such as manganese (II) chloride into polymer matrix could effectively promote the formation of char after the combustion process by reductive coupling mechanism [33]. The promotion of charring by manganese (II) chloride in TPS matrix could further improve the thermal stability by pre-dominating the chain fragmentation [33]. Thus, this also indicates that the increment of manganese (II) chloride loading level has significantly promoted the charring effect of TPS by increasing the formation of remaining char residues. The formation of char (remaining char residues) during combustion could cover on the surface of TPS samples and thus delay the thermal decomposition process [23].

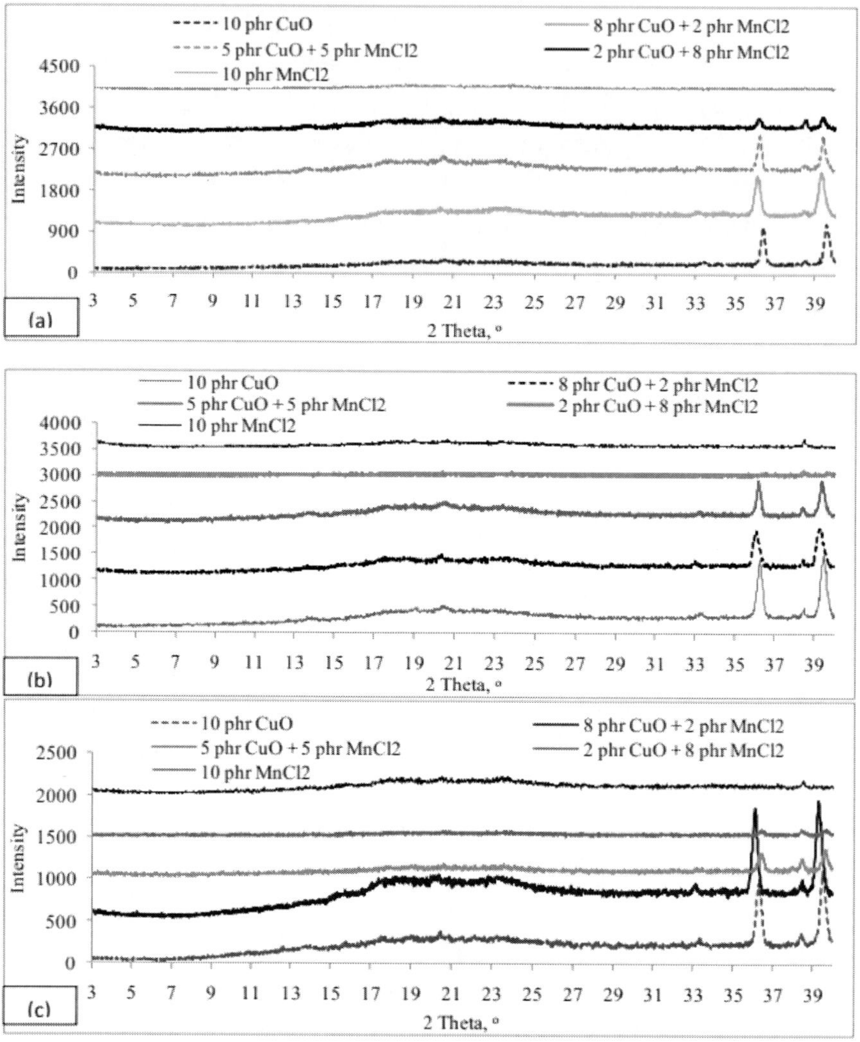

Figure 5. XRD pattern curves of thermoplastic starch added with various loading level ratio of copper (II) oxide/ manganese (II) chloride when subjected to irradiation dosages of (a) 0 kGy (un-irradiated), (b) 20 kGy and (c) 60 kGy.

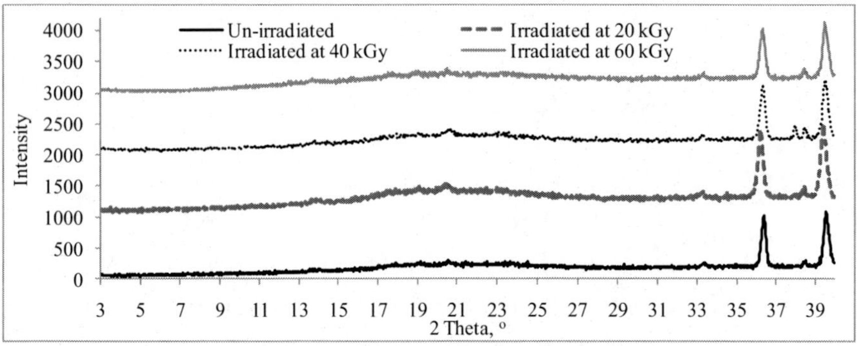

(a) 10 phr Copper (II) oxide

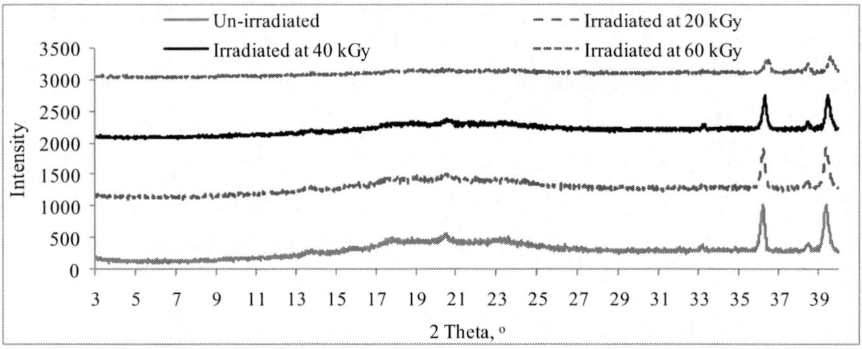

(b) 5 phr Copper (II) oxide + 5 phr Manganese (II) chloride

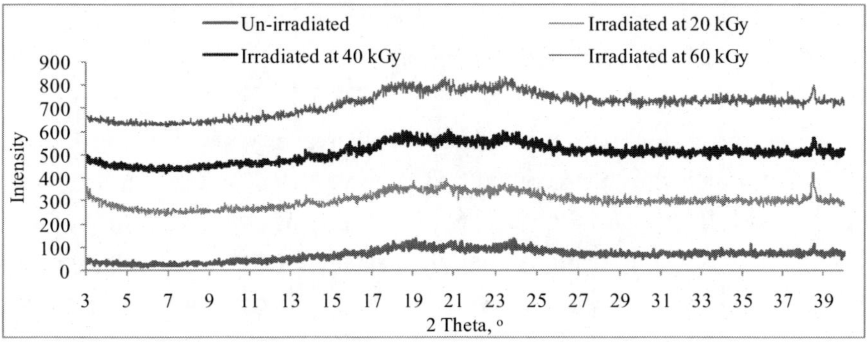

(c) 10 phr Manganese (II) chloride

Figure 6. XRD pattern curves of thermoplastic starch added with (a) 10 phr copper (II) oxide, (b) 5 phr copper (II) oxide and 5 phr manganese (II) chloride and (c) 10 phr manganese (II) chloride when subjected to various irradiation dosages.

X-Ray Diffraction Analysis

The XRD pattern curves of all un-irradiated and irradiated thermoplastic starch (TPS) added with different loading level ratios of $CuO/MnCl_2$ were illustrated in Figures 5 and 6. By referring to Figures 5 and 6, the appearance of two significant reflection peaks which are peak A and peak B can be observed to occur at $2\theta = 36.0^{\circ}$~36.4° and $2\theta = 39.2^{\circ}$~39.8° on the XRD curves of $CuO/MnCl_2$ added TPS samples. According to Figure 5(a), the peak A and peak B were observed to appear on the XRD curve of non-irradiated 10 phr copper (II) oxide added TPS matrix at $2\theta = 36.040^{\circ}$ and 39.555°, respectively. On the other hand, the peak A and peak B was found to be disappeared from XRD diffraction curve of 10 phr manganese (II) chloride added TPS. By referring to Figure 5(a), the increasing of manganese (II) chloride loading level in un-irradiated TPS samples was observed to gradually reduce the peak size of peak A and peak B. This indicates that the appearance of peak A and peak B on XRD curves of TPS matrix is attributable to the occurrence of copper (II) oxide particles. As the manganese (II) chloride increased, the crystallites of peak A and peak B in TPS matrix have been slightly reduced and ruptured by the manganese (II) chloride. Similar observation also can be found on all the irradiated $CuO/MnCl_2$ added TPS samples as shown in Figures 5(b) and 5(c).

Figure 6(a) illustrates the effect of increasing the electron beam irradiation dosages from 0 kGy to 60 kGy on the XRD pattern curves of 10 phr copper (II) oxide added TPS. By referring to Figure 6(a), the application of electron beam irradiation dosage up to 20 kGy was observed to slightly broaden the peak A and peak B of 10 phr copper (II) oxide added TPS samples. The peak broadening effect is attributed to the reduction in average crystallite size of peak A and peak B [27]. This is consistent with the calculated crystallite size for peak A and peak B as shown in Table 3 and 4. The crystallite sizes of peak A and peak B for 10 phr copper (II) oxide added TPS were observed to highly decrease from 341.6 Å and 320.7 Å to 289.8 Å and 269.9 Å, respectively when subjected to irradiation dosage up to 20 kGy. This also indicates that the application of low irradiation dosages could reduce the crystallite size by reducing the interaction between the original crystallite in copper (II) oxide and newly formed crystallite in amorphous region of thermoplastic starch matrix [27]. However, further increment in irradiation dosages up to 40 kGy has significantly reduced the peak width of peak A and peak B which indicating the increment of crystallite size of peak A and peak B as shown in Tables 3 and 4. This might be due to the higher crosslinking in TPS matrix could slightly enhance the interaction between original crystallite and the newly formed crystallite. By referring to Figure 6(b), the increasing of electron beam irradiation dosages from 0 kGy to 60 kGy was found to slightly reduce the size of peak A and peak B of 5 phr manganese (II) chloride + 5 phr copper (II) oxide added TPS samples. Besides, the peak A and peak B also observed to be broaden with increasing of irradiation dosages from 0 kGy to 60 kGy which indicating the reduction of crystallite size of peak A and peak B. The reduction of peak size after subjected to electron beam irradiation indicates the rupture of polymer chains in highly ordered arrangement [17]. Besides, the peak broadening of peak A and peak B also indicates that the rupture of newly formed crystallite in amorphous region which this could further reduce the interaction between original crystallite in copper (II) oxide and the newly formed crystallite [27]. For 10 phr manganese (II) chloride added TPS samples, the peak A and peak B are observed to be disappeared from the XRD pattern curves as shown in Figure 6(c). This shows that the appearance of peak A and peak B on the XRD pattern curves of TPS samples is mainly contributed to the occurrence of copper (II) oxide particles inside TPS matrix.

By referring to Figure 7, the increasing the manganese (II) chloride loading level (from 0 phr to 5 phr) has gradually decreased the crystallinity of all un-irradiated and irradiated TPS samples. This indicates that the addition of manganese (II) chloride particles into TPS matrix could disturb the highly ordered arrangement structure of polymer chains in TPS matrix. The disruption of highly ordered arrangement of polymer chains in TPS matrix into more random structure arrangement could lead to reduce the crystallitnity [34-35]. Besides, the increasing of manganese (II) chloride loading level also could highly induce the water vapour absorption ability of TPS matrix. The absorbed water vapour into polymer matrix also might slightly disturb the highly ordered arrangement structure in TPS matrix. However, further increment in manganese (II) chloride loading level from 5 phr to 8 phr has significantly increased the crystallinity of all un-irradiated and irradiated CuO/$MnCl_2$ added TPS samples except for 60 kGy irradiated samples. This might be due to the crystal structures presented in starch and copper (II) oxide particles could well interact with the crystal structure in manganese (II) chloride particles. Thus, the crystallinity of TPS matrix was significantly increased as the loading level of manganese (II) chloride increased from 5 phr to 8 phr. For all irradiated samples, the increasing of manganese (II) chloride loading level up to 10 phr has rapidly decreased the crystallinity of TPS matrix. This might be due to the absence of copper (II) oxide particles in TPS matrix did not contribute any increment effect in converging polymer chains in random arrangement structure into highly ordered arrangement structure.

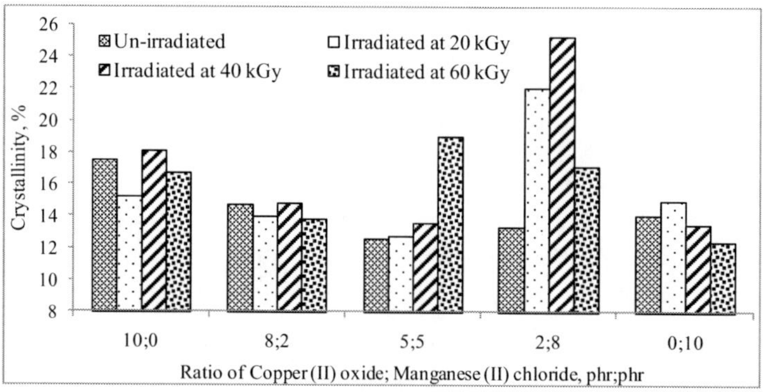

Figure 7. Effect of different loading level ratio of copper (II) oxide/manganese (II) chloride on the crystallinity of thermoplastic starch under various irradiation dosages.

For higher copper (II) oxide added TPS samples (10 phr copper (II) oxide and 8 phr CuO + 2 phr $MnCl_2$ added TPS samples), the application of irradiation dosage up to 20 kGy has significantly reduced the crystallinity as shown in Figure 7. This might be due to the low crosslinking formed in TPS matrix added with higher copper (II) oxide loading level has slightly ruptured the polymer chains which are highly ordered arranged in polymer matrix. The low crosslinking formed did not exhibit good interaction with the high copper (II) oxide particles inside the TPS matrix. On the other hand, the increasing of electron beam irradiation from 0 kGy to 60 kGy was found to gradually increase the crystallinity of TPS samples added with 5 phr CuO + 5 phr $MnCl_2$ and 2 phr CuO + 8 phr $MnCl_2$. This indicates that the increasing of electron beam irradiation dosages could induce the convergences of random polymer chains in amorphous region of TPS into higher ordered arrangement structures [27].

Table 3. The 2θ, crystallite size (L), d-spacing (d) and interchain separation (R) of deflection peak A on XRD curves of thermoplastic starch added with various loading level ratios of copper (II) oxide/manganese (II) chloride under irradiation dosages from 0 kGy to 60 kGy

Ratio of Copper (II) oxide/ Manganese (II) chloride, phr/phr	Irradiation dosage, kGy	Angle (2θ), o	Crystallite size (L), Å	d-spacing, Å	Interchain Separation (R), Å
10/0	0	36.040	341.6	2.489	3.111
	20	36.218	289.8	2.477	3.097
	40	36.297	302.9	2.472	3.090
	60	36.285	287.0	2.473	0.091
8/2	0	36.059	256.4	2.488	3.110
	20	36.063	252.9	2.488	3.109
	40	36.100	301.7	2.485	3.106
	60	36.075	296.3	2.487	3.108
5/5	0	36.154	359.7	2.481	3.102
	20	36.172	353.2	2.480	3.100
	40	36.257	354.6	2.475	3.093
	60	36.394	299.5	2.466	3.082
2/8	0	36.141	340.1	2.482	3.103
	20	36.440	464.1	2.463	3.078
	40	36.610	357.1	2.452	3.065
	60	36.412	287.2	2.465	3.081
0/10	0	-	-	-	-
	20	-	-	-	-
	40	-	-	-	-
	60	-	-	-	-

Table 4. The 2θ, crystallite size (L), d-spacing (d) and interchain separation (R) of deflection peak B on XRD curves of thermoplastic starch added with various loading level ratios of copper (II) oxide/manganese (II) chloride under irradiation dosages from 0 kGy to 60 kGy

Ratio of Copper (II) oxide/ Manganese (II) chloride, phr/phr	Irradiation dosage, kGy	Angle (2θ), o	Crystallite size (L), Å	d-spacing, Å	Interchain Separation (R), Å
10/0	0	39.555	320.7	2.276	2.845
	20	39.439	269.9	2.282	2.853
	40	39.508	293.9	2.278	2.848
	60	39.507	279.7	2.278	2.848
8/2	0	39.287	246.0	2.291	2.863
	20	39.281	229.3	2.291	2.864
	40	39.316	283.1	2.289	2.861
	60	39.290	287.6	2.290	2.863
5/5	0	39.361	333.2	2.286	2.858
	20	39.379	338.6	2.285	2.857
	40	39.462	343.3	2.281	2.851
	60	39.590	318.1	2.274	2.842
2/8	0	39.354	335.1	2.287	2.859
	20	39.670	360.7	2.269	2.837
	40	39.790	312.7	2.263	2.828
	60	39.614	261.7	2.272	2.840
0/10	0	-	-	-	-
	20	-	-	-	-
	40	-	-	-	-
	60	-	-	-	-

The d-spacing and interchain separation of deflection peak A and peak B for all un-irradiated and irradiated $CuO/MnCl_2$ added TPS samples are calculated and presented in Tables 3 and 4. By referring to Table 3, the increasing of manganese (II) chloride loading level from 0 phr to 10 phr was observed to pose insignificant effect in affecting the d-spacing and interchain of deflection peak A. This indicates that the addition of manganese (II) chloride particles did not provide significant effect in reducing or increasing the compactness of polymer chains arrangement in crystallite of peak A. In other words, the compactness of polymer chains arrangement inside crystallite of peak A did not significantly changed by the increasing of manganese (II) chloride loading level. However, the addition of 2 phr manganese (II) chloride was observed to slightly increase the d-spacing and interchain separation of peak B from 2.276 Å and 2.845 Å to 2.291 Å and 2.863 Å, respectively. This also indicates that the occurrence of manganese (II) chloride particles in TPS matrix could cause the polymer chains in crystallite of deflection peak B to be less closely arranged and compacted [35]. This might be due to the manganese (II) chloride particles in TPS matrix tended to absorb water vapor from surrounding. The water absorption by manganese (II) chloride particles into TPS matrix could reduce the compactness of polymer chains arrangement in crystallite of deflection peak B by occupying the spaces between the polymer chains in crystallite (peak B). Thus, the d-spacing and interchain have been significantly increased. By referring to Table 4, further increment of manganese (II) chloride loading level in thermoplastic starch matrix has slightly reduced the d-spacing and interchain separation of polymer chains in crystallite of peak B. This might be due to further absorption of water vapor into TPS matrix could cause the polymer chains in crystallite structure to be more closely packed or compacted [35].

According to Tables 3 and 4, the increasing of irradiation dosages from 0 kGy to 60 kGy on all the CuO/$MnCl_2$ added TPS samples has gradually reduced the d-spacing and interchain separation of crystallites for peak A and peak B. This could be due to the crosslinking induced by electron beam irradiation could tighten the polymer chains arranged inside TPS matrix to be more closely packed and compacted. The spacing distance between the polymer chains inside crystallites of peak A and peak B has been further reduced by the tightening effect of crosslinking networks formed in polymer matrix [35]. Thus, the compactness of crystallites in peak A and peak B has been induced by the crosslinking formed in polymer matrix. However, the d-spacing and interchain separation of crystallite in peak B of 10 phr copper (II) oxide added thermoplastic starch were found to slightly increase when irradiated to 20 kGy. The application of low irradiation dosages ($\leq$ 20 kGy) could slightly improve the dispersion of copper (II) oxide particles in thermoplastic starch matrix. The well dispersion of copper (II) oxide particles in TPS matrix could further increase the spacing distance between the molecular chains arranged in crystallites of deflection peak B. Thus, the d-spacing and interchain separation for crystallites of peak B has been increased. Further increment in irradiation dosages from 20 kGy to 60 kGy has slightly decreased the d-spacing and interchain separation of crystallites in peak B for 10 phr copper (II) oxide added TPS. This is due to the tightening effect of crosslinking formed in polymer matrix in linking the polymer chains together. This caused the polymer chains in crystallites of peak B to be more closely arranged and compacted.

CONCLUSION

This study has conducted an investigation on the effect of adding $CuO/MnCl_2$ in different loading level ratios on physico-mechanical and thermal properties of electron beam irradiated TPS samples. The tensile strength and Young's modulus of un-irradiated and all irradiated $CuO/MnCl_2$ added TPS samples have gradually increased as the loading level of manganese (II) chloride added into TPS matrix was gradually increased. This is due to the manganese (II) chloride particles could form a strong interaction with starch molecules and further induce the required tensile stress to break the TPS samples. Subsequently, the tensile strength and Young's modulus of TPS matrix have significantly been increased. On the other hand, the application of electron beam irradiation was found to gradually decrease the tensile strength and Young's modulus of all $CuO/MnCl_2$ added TPS samples. This is due to the electron beam irradiation could induce the formation of intermolecular linkages between hydroxyl groups (or crosslinking) in TPS matrix. The formation of intermolecular linkages between hydroxyl groups in TPS matrix could further weaken the intramolecular and intermolecular bonding inside TPS matrix and thus also decrease the resistance ability of polymer matrix towards the applied straining stress. The increasing of copper (II) oxide loading level has gradually increased the thermal degradation temperatures ($T_{5\%}$, $T_{15\%}$, $T_{25\%}$, $T_{35\%}$, $T_{45\%}$ and $T_{55\%}$) of TPS samples. This is attributed to the copper (II) oxide particles could delay the thermal degradation of TPS samples by acting as inorganic catalysts during combustion process. Thus, the increasing of copper (II) oxide in TPS matrix could enhance the thermal stability of TPS samples. The increasing of irradiation dosages has slightly delayed the thermal degradation temperatures ($T_{25\%}$, $T_{35\%}$, $T_{45\%}$ and $T_{55\%}$) to higher thermal decomposition temperature due to the formation of crosslinking networks. The increasing of manganese (II) chloride loading level in un-irradiated and all irradiated TPS matrix has gradually decreased the crystallinity of TPS samples. This due to the manganese (II) chloride particles could disturb the highly ordered arrangement structure of polymer chains in TPS matrix and converge into more random structure arrangement. For 5 phr CuO + 5 phr $MnCl_2$ and 2 phr CuO + 8 phr $MnCl_2$ added TPS samples, the increasing of electron beam irradiation dosage has gradually increased the crystallinity of TPA matrix. This is due to the formation of crosslinking in TPS matrix could induce the convergences of random polymer chains in amorphous region of TPS to arrange into higher ordered arrangement structures.

ACKNOWLEDGMENTS

The authors are very appreciating with the kindliness of Malaysian Nuclear Agency, Bangi, Selangor for allowing usage of their equipments on performing this research.

REFERENCES

[1] K. M. Dang, R, Yoksan. 115 *Carbohydr. Polym.* (2015) 575.

[2] E. M. Teixeira, A. A. S. Curvelo, A. C. Corrêa, J. M. Marconcini, G. M. Glen, L. H. C. Mattosa, L. H. C. 37 *Ind. Crops Prod.* (2012) 61.

[3] J. P. López, P. Mutjé, A. J. F. Carvalho, A. A. S. Curvelo, J. Gironès. 44 *Ind. Crops Prod.* (2013) 300.

[4] J. Cai, Z. Xiong, M. Zhou, J. Tan, F. Zeng, M. Ma, S. Lin, H. Xiong. 102 *Carbohydr. Polym.* (2014) 746.

[5] M. B. K. Niazi, M. Zijlstra, A. A. Brockhuis. 49 *Eur. Polym. J.* (2013) 1861.

[6] H. Schmitt, A. Guidez, K. Prashantha, J. Soulestin, M. F. Lacrampe, P. Krawczak. 115 *Carbohydr. Polym.* (2015) 361.

[7] M. Mondragón, E. M. Hernández, J. L. Rivera-Armentam, F. J. Rodríguez-González. 77 *Carbohydr. Polym.* (2009). 80.

[8] L. T. Sin, W. A. W. A. Rahman, A. R. Rahmat, T. T. Tee, S. T. Bee, C. Y. Low. 14 *J. Manufac. Proc.* (2012) 8.

[9] K. Frost, E. Johansson, R. Shanks, S. Goto, G. Kirwan, G. 32 *Polym. Test.* (2013) 343.

[10] H. Abdillahi, E. Chabrat, A. Rouilly, L. Rigal. 50 *Ind. Crops Prods.* (2013) 104.

[11] L. Castillo, O. López, C. López, N. Zaritzky, M. A. Carcía, S. Barbosa, M. Villar, M. 95 *Carbohydr. Polym.* (2013) 664.

[12] T. T. Tee, L. T. Sin, R. Gobinath, S. T. Bee, D. Hui, A. R. Rahmat, I. Kong, Q. Fang. 47 *Compos Part B: Eng.* (2013) 238.

[13] S. T. Bee, C. T. Ratnam, L. T. Sin, T. T. Tee, D. Hui, A. A. H. Kadhum, A. R. Rahmat, J. Lau. 63 *Compos. Part B: Eng.* 2014, 141.

[14] S. Dalai, C. Wenxiu. Radiation effects on LDPE/EVA blends. 8 6 *J. Appl. Polym. Sci.* (2002) 1296.

[15] M. E. Martinez-Pardo, M. R. Zuazua, V. Hernández-Mendoza, J. Cardoso, R. Montiel, H. Vázquez H. 105 *Nucl. Instrum. Meth. Phys. Res. B* (1995) 258.

[16] A. Zezin, V. I. Feldman, N. A. Shmakova, S. P. Valueva, V. K. Ivanchenko, N. I. Nikanorova. 265 *Nucl. Instrum. Meth. Phys. Res. B* (2007) 334.

[17] X. Jiang, T. Jiang, L. Gan, X. Zhang, H. Dai, X. Zhang, X. 90 *Carbohydrate Polymers* (2012) 1677.

[18] J. Prachayawarakorn, W. Pomdage, W. 61 *Mater. Des.* (2014) 264.

[19] R. Wongsagonsup, T. Pujchakarn, S. Jitrakbumrung, W. Chaiwat, A. Fuongfuchat, S. Varavinit, S. Dangtip, M. Suphantharika, M. 101 *Carbohydr. Polym.* (2014) 656.

[20] S. T. Bee, A. Hassan, C. T. Ratnam, T. T. Tee, L. T. Sin, L. T. 33 *Polymer Composites* (2012) 1883.

[21] S. M. Grimes, H. Lateef, A. J. Jafari, L. Mehta. 91 *Polym. Degrad. Stab.* (2006) 3274.

[22] A. Basfar. (2002). 63 *Radiat. Phys. Chem.* (2002) 505.

[23] S. T. Bee, C. T. Ratnam, L. T. Sin, T. T. Tee, W. K. Wong, J. X. Lee, A. R. Rahmat, A. R. (2014). 334 *Nucl. Instrum. Meth. Phys. Res. B* (2014) 18.

[24] Ng, H. M.; Bee, S. T.; Ratnam, C. T.; Sin, L. T.; Phang, Y. Y.; Tee, T. T. *Nucl Inst Meth Phys Res* B 2014, 319, 62.

[25] S. T. Bee, A. Hassan, C. T. Ratnam, T. T. Tee, L. T. Sin, L. T. 299 *Nucl. Instrum. Meth. Phys. Res. B* (2013) 42.

[26] Z. Osawa. Role of Metals and Metal-Deactivators in Polymer Degradation. 20 *Polym. Degrad. Stab.* (1988) 203.

[27] V. Kumar, Y. Ali, R. G. Sonkawade, A. S. Dhaliwal, A. S. 287 *Nucl. Instrum. Meth. Phys. Res. B* (2012) 10.

[28] F. J. Rodriguez-Gonzalez, B. A. Ramsay, B. D. Favis. 58 *Carbohydr. Polym.* (2004) 139.

[29] V. G. Kumar, M. V. Bhandary, A. N. Bhat. (1989). 146 *Thermochim. Act.* (1989) 2011.

[30] K. Kaewtatip, J. Thongmee. 45 *Mat. Des.* (2013) 586.

[31] F. Holdsworth, A. R. Horrocks, B. K. Kandola, D. Price, D. 110 *Polym. Degrad. Stab.* (2014) 290.

[32] S. G. Hong, C. S. Yeh. 83 *Polym. Degrad. Stab.* (2004) 529.

[33] J. Jang, J. Kim, J. Y. Bae. 88 *Polym. Degrad. Stab.* (2005) 324.

[34] M. E. Baker, A. H. Windle, A. H.. 42 *Polym.* (2001) 667.

[35] S. Bhadra, D. Khastgir. 27 *Polym. Test.* (2008) 851.

In: Polymer-Matrix Composites ISBN: 978-1-53613-398-1
Editors: Elias Rice and Beth Sparks

Chapter 5

ANALYSIS OF STARCH ADDED POLYETHYLENE BLENDS CHARACTERISTICS WHEN SUBJECTED TO ACID AND ALKALINE ATTACKS

Ang Rui-Ren, Lee Tin Sin[1,*], Bee Soo-Tueen[1,†], Tiam-Ting Tee[1], A. R. Rahmat[2] and Wei-Long Wong[1]

[1]Department of Chemical Engineering,
Lee Kong Chian Faculty of Engineering and Science,
Universiti Tunku Abdul Rahman, Bandar Sungai Long,
Jalan Sungai Long, Selangor, Malaysia
[2]Department of Polymer Engineering,
Faculty of Chemical Engineering,
Universiti Teknologi Malaysia, Johor, Malaysia

[*] Corresponding Author: Email: direct.tinsin@gmail.com; Tel: +60 3 9086 0288; Fax: +603 9019 1989.
[†] Corresponding Author Email: direct.beest@gmail.com.

Abstract

This chapter aims at investigating the physico-mechanical characteristic of low density polyethylene (LDPE) and starch blends when undergone acid and alkaline attacks. It was found that the mechanical properties of the polymer blends have improved at the initial stage. This is because the acid solution (hydrochloric acid) and the alkaline solution (sodium hydroxide) consumed the starch portion enables the remaining LDPE chains to rearrange in highly consistent structure without the disruption of starch molecules. However, as the duration of exposure increased, the polymer blends undergone degradation and thus resulted in poor tensile properties. It was also found that the X-ray diffraction peaks become broader at initial treatment stage indicating formation of crystallinity structure. Under prolong exposure to acid and alkaline conditions whereby starch has been fully consumed, the acid and alkaline components attacked the LDPE bonds directly. This can be evidenced from the electron microscope observation of polymer blends after long period of exposure where the matrix showed discontinuity along with voids with short and thick fibrils.

Keywords: starch, polyethylene, acid/alkaline attack, degradation

1. Introduction

The researches about the blending of polymer and agricultural compounds have been widely reported since decades ago [1]. The reasons of blending of agricultural compounds are mainly because of agricultural residues are available abundantly at cheap cost as well as its renewable characteristic for sustainable supply [2]. Furthermore, most of the agricultural residue is a natural resource which is considered to be harmless to environment when being disposed in environment [3]. Subsequently, there are demands on starch based polymer compound for consumer disposable food packaging and food utensils [4]. This innovation is typically well known to be practicable especially blending of agricultural residue with biodegradable polymer [5]. In fact the famous starch based polymer compound- MaterBi® is belong to this kind of material which has been well accepted by consumers [6]. On the other hand, another type of works that involves blending of agricultural ingredients with non-degradable polymer likes wood flour with polyethylene where the wood flour is used as filler [7]. Similar aspect also applied to the starch based polyethylene compound i to promote degradation of polymer where the starch portion can be easily consumed by microorganism when disposed to the environment.

Many studies have been performed to evaluate the physico-mechanical properties of polyethylene-starch blends and correlate with its biodegradability. For instance, Huang, Roan, Kuo and Lu [8] found that the both maleic anhydride grafted LDPE (low density polyethylene) and acrylic acid grafted LDPE added polyethylene-starch blends possessed 65 wt.% weight loss after two weeks of biodegradation. Greizerstein, Syracuse and Kostyniak [9] reported that polyethylene-starch leaf bags showed oxidation, loss of starch and embrittleness after 14 days soil burial. The similar observation was also found by Danjaji, Nawang, Ishiaku, Ismail and Ishak [10] that exposure of the polyethylene-starch to weathering tend to cause massive deterioration such as turbidity, discoloration, embrittleness and dimensional changes. Throughout the soil burial period, the presence of holes on the samples can be observed which indicates occurrence of biodegradability activity. Arvanitoyannis, Biliaderis, Ogawa and Kawaski [11] reported that the rate of biodegradability was significant when the starch content exceeded 10% (w/w). However, Prachaywarakorn, Sangnitidej and Boonpasith [12] discovered that the biodegradation weight loss was lesser for polyethylene-starch blends added with vinyltrimethoxysilane compatibilizers as resulted by the improved phase compatibility between the starch and LDPE.

As can be noticed from most of the researches done previously [13], there is lack of information to describe the degradation of polyethylene-starch compound when exposing to acid and alkaline condition. In fact, this information is useful to polyethylene-starch compound to be used for food packaging application which is highly potential to contact with acid and alkaline content [14]. The pH condition would definitely cause some extent of degradation when being used for food packaging before consumption. Thus in this study, the mechanical properties and morphology-structural of the LDPE-starch polymer blend corresponds to different concentration of acid and alkaline solution are reported. These results can also be used to provide additional information for the food industries for selection of appropriate packaging system to maintain the freshness of foods as well as prolong storage period.

2. METHODOLOGY

2.1. Materials

Low Density Polyethylene (LDPE) was used as polymer resin in this study with grade LDF200GG of melt flow index 2.0 g/10min produced by *Lotte Chemical Titan.*, Malaysia. Native tapioca starch was purchased from Thye Huat Chan Sdn. Bhd., Malaysia and was used as received. Meanwhile the hydrochloric acid (HCl)) and sodium hydroxide (NaOH) used was supplied by R&M Chemicals with hydrolysis percentage of 87-89 mol%.

2.2. Samples Preparation

The samples were prepared by blending LDPE with various compositions 5-20 parts per hundred resin (phr) of starch using twin screw extruder Nanjing Giant SHJ-20 at 7 rpm with all zones set at 190°C. The extrudate was cooled in water and fed into a pelletizer to obtain polymer pellets. The pellets were then dried at 60°C for 24 hours. The polymer pellets were compression moulded by using hot press machine under a heating temperature of 160°C to form polymer sheets. The polymer sheets were cut into dumbbell-shaped polymer samples in accordance to ASTM D1822 standard. The dumbbell-shaped polymer samples were immersed in acid and alkali with concentration of 1 M and 3 M at room temperature of 25°C. The polymer samples for each formulation were collected once every 3 days until 15 days for various analyses as described below.

2.3. Tensile Test

Tensile test was conducted to evaluate the tensile strength, Young's modulus of elasticity, and elongation at break of the polymer samples. The samples were cut into dumbbell-shaped in accordance to ASTM D1822 standard and tested using Instron Universal Testing Machine (Model 4302 Series IX). The tensile test was carried out at room temperature (25°C) using 2 kN load cell with crosshead speed of 50 mm/min. The final value of tensile property data for each sample was obtained from the average of five specimens.

2.4. X-Ray Diffraction Analysis (XRD)

The changes of the acid and alkaline degraded LDPE-starch compound were analyzed by using the XRD-6000 Shimadzu X-ray Diffractometer at 40 kV and 40 mA with the scanning rate of 1°/min. The samples were scanned over a range of $2\theta = 0$ to 40°. The crystallite sizes of the samples were calculated by using the following formula (Scherrer formula):

$$L = \frac{k\lambda}{b\cos\theta} \tag{1}$$

where L = crystallite size, Å; k = Scherrer constant; b = FWHM, radian

2.5. Scanning Electron Microscopy (SEM)

The surface morphologies of fractured samples were investigated by using a SEM machine of Hitachi S3400N with a voltage of 15 kV. The fractured surfaces of the samples were observed under the microscopy at magnification of 1000×, 8000× and 15000×.

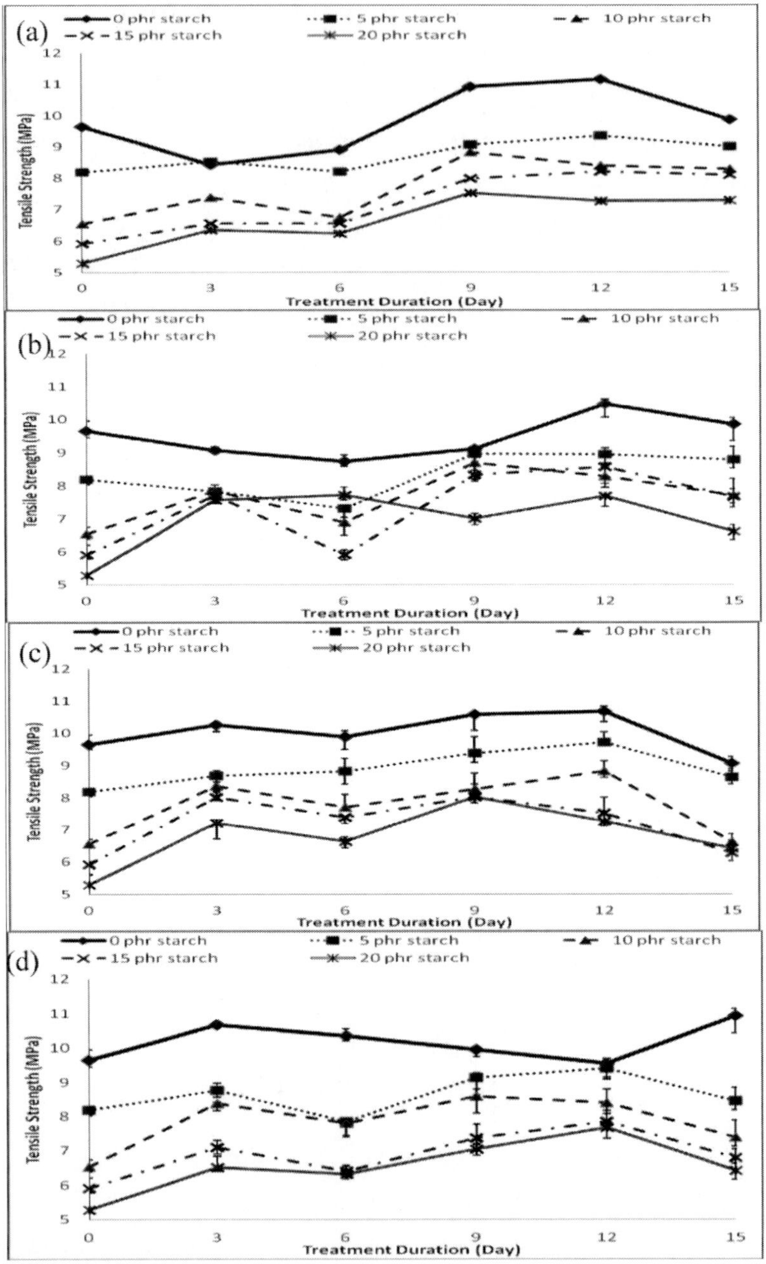

Figure 1. Tensile strength of LDPE-starch blend treated/immersed at (a) 1M HCl solution, (b) 3M HCl solution, (c) 1M NaOH solution and (d) 3M NaOH solution for 15 days.

3. Results and Discussion

3.1. Tensile Properties

Figure 1 shows the tensile strength of LDPE-starch polymer blend when treated in 1M and 3M HCl acid until 15 days. It can be found that before acid and alkaline attacks, pristine LDPE exhibited the highest tensile strength 9.65 MPa. When merely adding of 5 phr of starch into LDPE, the tensile strength has drastically dropped to 8.81 MPa. This was followed by reduction of tensile strength to 6.54 MPa, 5.91 MPa and 5.29 MPa for 10, 15 and 20 phr of starch, respectively. This indicates that as more starch was added in the polymer blend, the mechanical properties declined. Such phenomenon may due to the presence of starch has affected the intermolecular force within the blends. Since starch and LDPE are different in polarity, they are deemed to be incompatible with each other [11, 15]. In another word, when both of them were blended together, a very poor interfacial adhesion can occur between the phases [16]. As a result, the tensile strength of the LDPE-starch polymer blend decreased drastically as the composition of the starch increased.

When LDPE-starch blends were immersed into 1M HCl solution, it can be found that all the compositions exhibited common trends, where the tensile strength increased initially, i.e., 3rd and 6th day and decreased afterwards. The tensile strengths were further increasing, i.e., 9th and 12th day and followed by declining again. When the LDPE-starch polymer blend was immersed into 1M HCl solution, the H^+ ions attack the starch instead of LDPE. This is mainly because starch is carbohydrate consisting of a large number of glucose units joined by glycosidic bond. The presence of oxygen rich functional group such as C–OH and C–O–C is susceptible to acid reaction as compared to high inertness of C–H only in LDPE [15]. In other words, when there is a close match of polarities of the components of reactions, the solvent (i.e., HCl) tends to attack the starch portion [17]. When the starch molecules with irregular shapes and size have been consumed, the polymer chains are able to align better when subjected to extension and thus increased the tensile strength. Meanwhile, as the days of immersion in HCl solution increased, the attacked starch would further forming free radicals resulted from the side-chain reaction of the H^+ ions. The formed free radicals then attacked the LDPE backbone and breakdown the chains into smaller ones. Such phenomenon is known as chain scissioning [17, 18]. This has caused the reduction of thc molecular weight and lowered the mechanical strength of the polymer blends. As more free radicals were being generated, there was a high possibility that the random chemical reactions took place which resulted in cross-linking and thus increased the tensile strength of the polymer blends (i.e., higher tensile strength at 9th – 12th days) [19-21]. Nevertheless, the effect of chains scissoning is greater than the effect of cross-linking, which resulted inferiority in the tensile strength.

On the other hand, when LDPE-starch blends were immersed into 3M HCl, it can be observed that the result showed similar trends compared to those which were treated using 1M HCl solution in a milder condition. By comparing the results obtained on the 3rd day, it can be observed that most of the polymer blends, except of the 5 phr of starch, experienced higher increment in tensile strength. This was mainly due to higher concentration of HCl solution which was being used. It can be deduced that when higher concentration of HCl solution was being used, the rate of starch being attacked by the H^+ ions become pronounce, which resulted in faster realignment of polymer chains and increased in tensile strength. However such phenomenon was not found for polymer blend with 5 phr of starch. This may due to low amount of starch being added into the polymer blend. Thus, the effect of starch attacked by the H^+ ions was not as significant as compared to the others. Besides that, all of the polymer blends showed lower tensile strength compared to those which were immersed into 1M HCl. It was deduced that when higher concentration of HCl was being used, the amount of H^+ ions attacking the backbone of the polymer should be greater and thus resulted in poorer tensile strength [17].

In addition, when LDPE-starch blends were immersed into 1M NaOH solution, it can be found that all the compositions showed common trends, where the tensile strength increased initially, i.e., 3rd and 6th day and decreased later, i.e., 6th day. Meanwhile, the tensile strength increased on 9th and 12th day followed by declining afterward. When the LDPE-starch polymer blend was immersed into NaOH solution, the OH^- ions present in the NaOH solution dissolved the starch molecules [17]. As the starch molecules with irregular shapes and size have been dissolved, the polymer chains became free from disruption caused by incompatibility starch and thus improved the overall tensile strength of the polymer blend. Similarly, as the days of immersion increased, the attacked starch molecules formed free radicals which then attacked the LDPE backbone and caused chain scissioning of the LDPE chain [18]. This lowered the molecular weight of the polymer blends and deteriorated its mechanical strength [17]. When the attack of OH^- ions on the starch molecules generated substantial amount free radicals, random chemical reactions may be taken place and caused crosslinking in the polymer blend [19-21]. This subsequently enhanced the tensile strength of the polymer blends. Importantly, the effect of chains scissoning was greater than the effect of crosslinking, subsequently still causing reduction of the tensile strength.

Meanwhile, when the LDPE-starch blends were exposed to 3M NaOH solution, it was anticipated to observe similar trends of tensile strength with faster rate as in the treatment using 3M HCl solution. However, they showed similar trends compared to those treated with 1M NaOH solution. The only difference observed was that during the 6th day of treatment, most of the polymer blends treated in 3M NaOH solution experiences greater loss of tensile strength. As higher concentration of NaOH solution was used, there was more OH^- ion attacking the starch molecules in the polymer blends. This increased the number of free radicals attacking the LDPE backbone, and further decreased the tensile strength of the polymer blends [17, 18]. However, the tensile strength later bounced back to the values similar to those exposed to 1M NaOH solution.

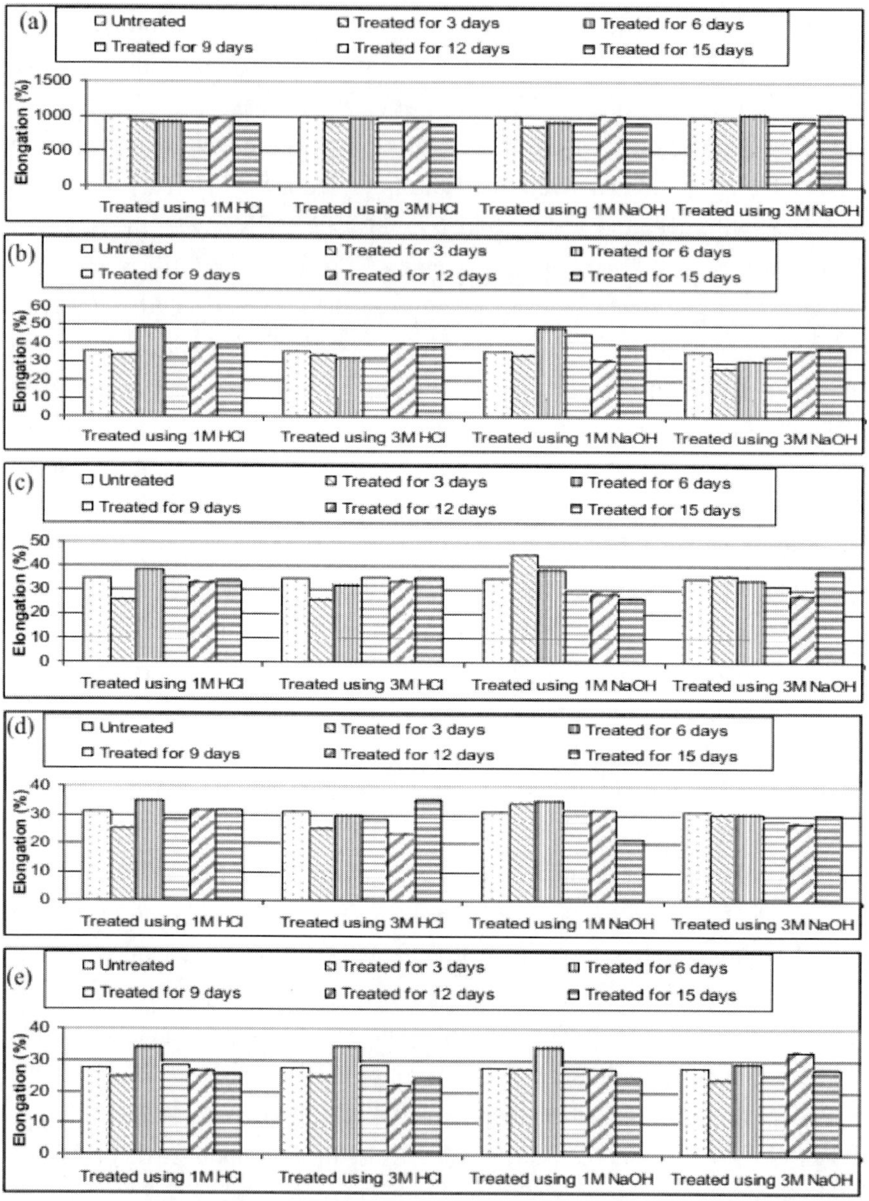

Figure 2. Tensile elongation of (a) pure LDPE, (b) 5 phr starch added polymer blend, (c) 10 phr starch added polymer blend, (d) 15 phr starch added polymer blend, and (e) 20 phr starch added polymer blend treated/immersed at 1M HCl solution, 3M HCl solution, 1M NaOH solution and 3M NaOH solution for 15 days.

Figure 2 shows the elongation of LDPE-starch polymer blend when immersed in HCl and NaOH solution until 15 days. It was found that prior to occurrence of acid and alkali attacks, the pure LDPE exhibited the highest elongation at break of 99.13%. As 5 phr of starch was blended into LDPE, the elongation at break has drastically dropped to 36.02%. This was followed by reduction of elongation at break to 34.57%, 31.04% and 27.53% for 10, 15 and 20 phr of starch, respectively. According to St-Pierre et al. [16], the elongation at break is highly sensitive to the interfacial interaction between the phases of the blend. Due to the poor matrix/load interface between the LDPE and starch, the addition of starch into the LDPE produced general trend of filler affected polymer properties, where the elongation decreased as the starch content increased [16, 22]. In addition, when the polymer blends were treated using 1M HCl, it can be found that the elongation at break of the polymer blends decreased. This was probably due to degradation induced by radical chain reactions which resulted in crosslinking and branching along the polymer chains [17]. On the other hand, when the LDPE-starch blends were treated in 3M HCl until 15 days, they exhibited similar trends but with obvious lower tensile elongation. When higher concentration of acid solution was being used, there were H^+ ions attacking the polymer blends, which resulted in more radicals being produced. This caused the polymer blends to experience greater degree of crosslinking within the matrix [19-21]. Thus, the tensile elongation was reduced. Similarly, when the polymer blends were immersed in 1M NaOH solution and 3M NaOH solution, elongation at break of the polymer blends decreased as well. Such trends may be because of the polymer degradation induced by radical chain reactions which resulted in crosslinking and branching along the polymer chains [17, 18].

3.2. X-Ray Diffraction (XRD) Analysis

The XRD pattern of LDPE-starch blends added with increasing starch loading level is illustrated in Figure 3(a). Two significant diffraction peaks (peak A and peak B) appear on XRD curves for all LDPE-starch blends at $2\theta \approx 22.1^{\circ}$ and 24.5°, respectively as shown in Figure 3(a). By referring to Figure 3(a), it can be observed that the diffraction peak width is in relation to the composition of starch. This is mainly because when the added starch tended to disrupt the arrangement of the polymer chains which resulted in a more randomized order polymer chain arrangement [23]. Furthermore, this also resulted in the peak broadening effect. The observation results on peak intensity were in agreement with the crystallinity results obtained from XRD analysis. By referring to Figure 4, the crystallinity of LDPE-starch blends was gradually decreased from 17.4% to 17.0% with increasing of starch loading levels from 0 phr to 20 phr. The decrement in crystallinity might be due to the addition of starch into LDPE matrix could disrupt the highly ordered arrangement structures in LDPE matrix by destructing the highly ordered structures in polymer matrix into randomized ordered arrangement. Thus, the addition of starch into LDPE matrix would cause the convergence of highly ordered of polymer chains arrangement towards a more randomized order polymer chains arrangement [23]. On the other hand, the reduction in crystallinity also might be due to the incompatibility of LDPE and starch polymers where LDPE is non-polar polymer while starch is a polar polymer [24].

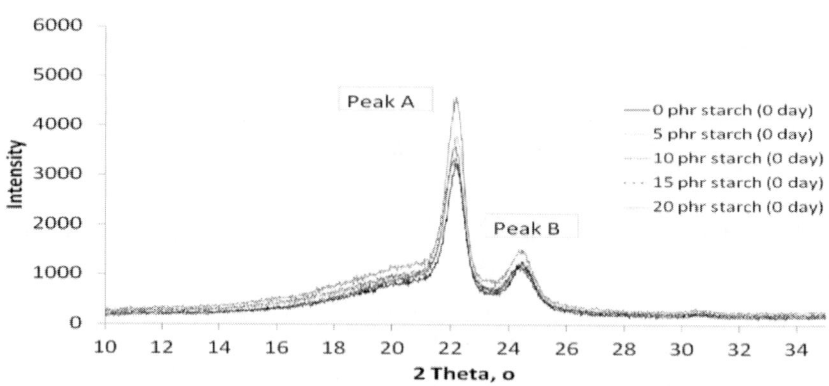

Figure 3. (Continued).

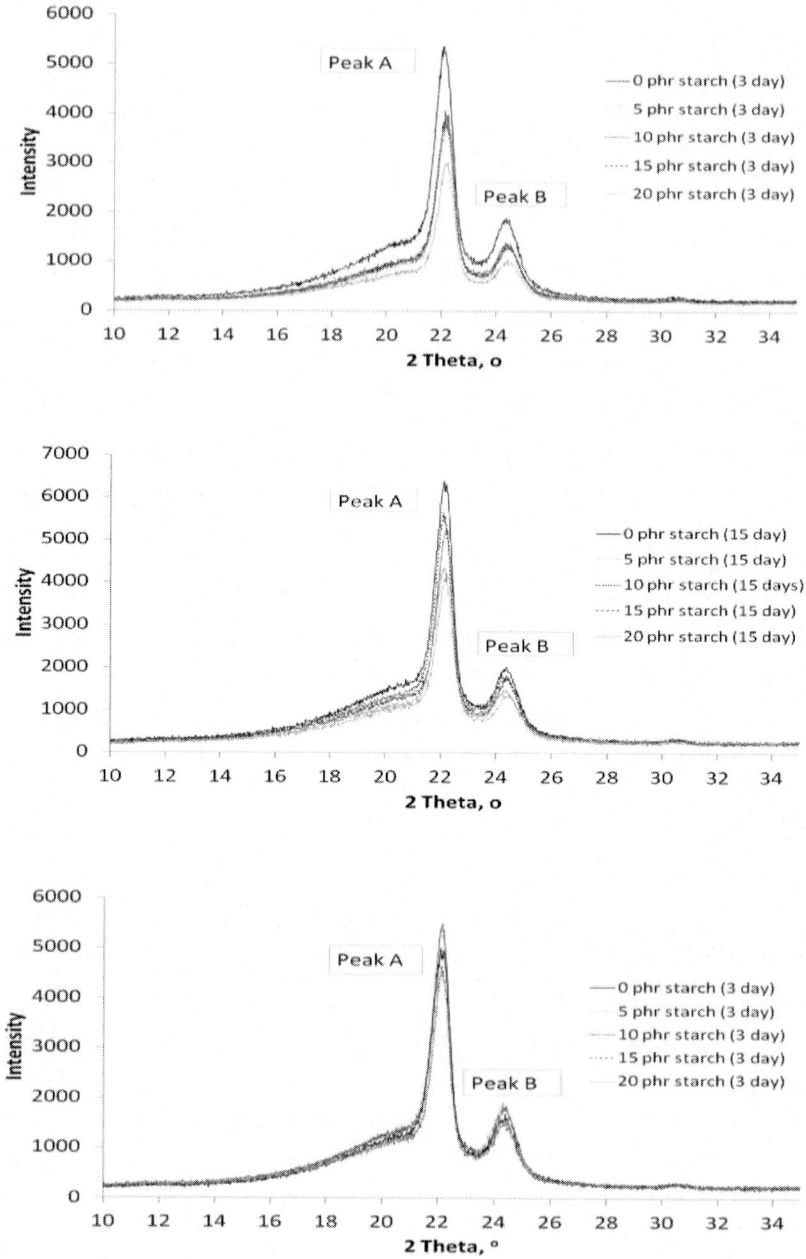

Figure 3. (Continued).

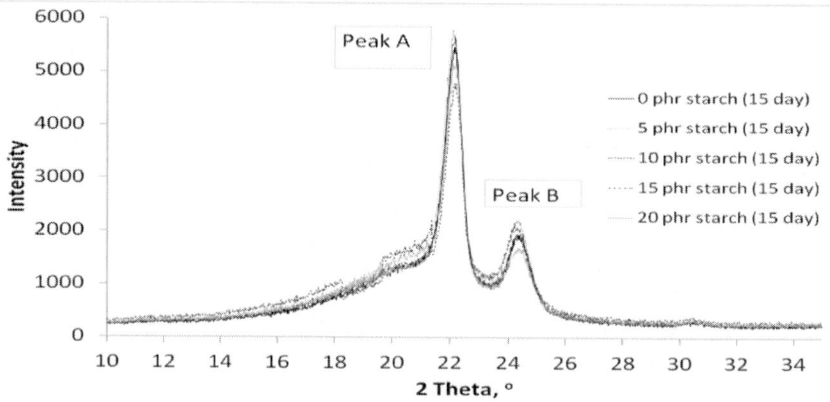

Figure 3. XRD pattern of (a) un-aged LDPE-starch blends with increasing starch percentages, (b) LDPE-starch blends treated/immersed for 3 days in 3M HCl solution, (c) LDPE-starch blends treated/immersed for 15 days in 3M HCl solution, (d) LDPE-starch blends treated/immersed for 3 days in 3M NaOH solution, and (e) LDPE-starch blends treated/immersed for 15 days in 3M NaOH solution.

Two significant diffraction peaks of peak A and peak B were also observed on XRD pattern curves of all the samples immersed in 3M HCl solution at $2\theta \approx 22.1^{o}$ and 24.5^{o}, respectively as shown in Figure 3(b) and Figure 3(c). By referring to both figures, it was found that the diffraction peaks become broader as they were being exposed to 3M HCl solution which indicated lowering in their crystallinity. Meanwhile, it can be observed that the peak width of the pure LDPE was the broadest among all polymer blends. This was caused by the absence of starch molecules in the polymer blend. As discussed earlier, the extent of dissolution between the solvents and the polymer depends on their solubility parameters or polarities [17]. Therefore, since there were no starch molecules in the pristine LDPE, instead of attacking the starch, the H^{+} ions from HCl attacked the LDPE bond directly. This caused chain scissioning of the LDPE polymer chains where its crystal structure became lack of rigidity and tends to rupture easily.

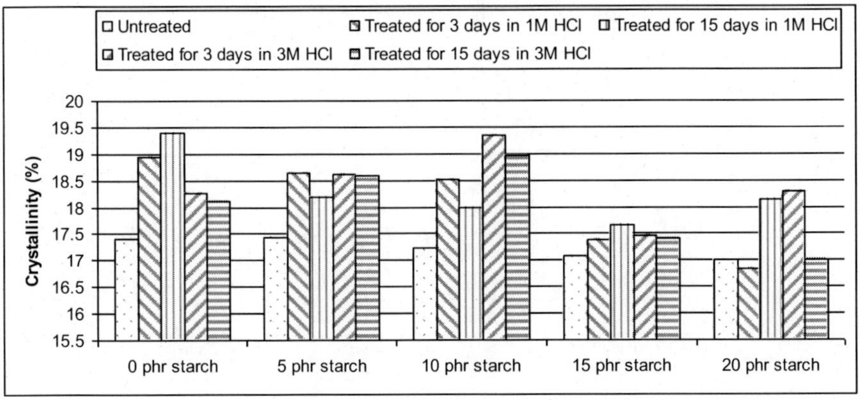

(a)

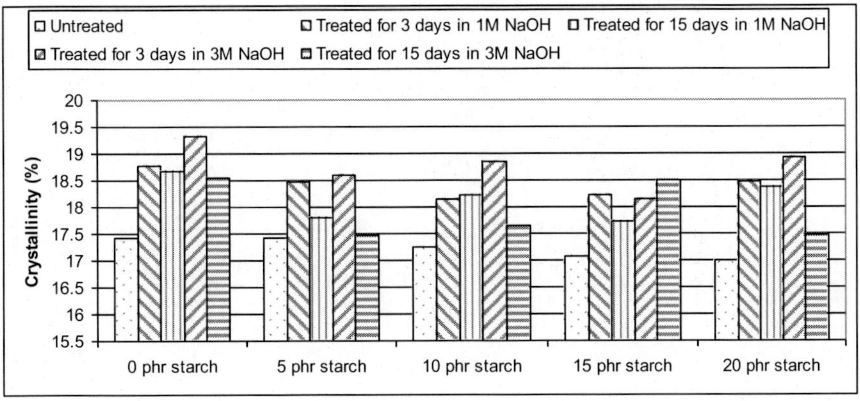

(b)

Figure 4. Crystallinity of LDPE-starch blends treated/immersed with (a) 1M HCl solution and 3M HCl solution, and (b) 1M NaOH solution and 3M NaOH solution.

On the other hand, the peak A and peak B at $2\theta \approx 22.1$ ° and 24.5 ° were also observed when samples were immersed in 3M NaOH as shown in Figure 3(d) and Figure 3(e). Similarly, it was observed that the diffraction peaks become broader as they were degraded in 3M NaOH solution. This indicates decrement of their crystallinity. Again, when the LDPE-starch polymer blend was immersed into NaOH solution, the OH^- ions present in the NaOH solution attack the starch molecules and caused chain scissioning of the LDPE polymer chains [17].

Table 1. The 2Θ and crystallite size of LDPE-starch blends when treated/immersed in 1M HCl solution for 3 days and 15 days

Loading level of starch, phr	1M HCl treated/immersed duration (day)					
	0 day		3 days		15 days	
	Peak A					
	2Θ, °	Crystallite size (L), Å	2Θ, °	Crystallite size (L), Å	2Θ, °	Crystallite size (L), Å
0	22.1263	105.69	22.0965	108.25	22.0441	103.48
5	22.1323	102.93	22.1188	106.73	22.0642	103.26
10	22.079	102.53	22.0906	107.52	22.0831	103.67
15	22.0328	101.75	22.0587	108.92	22.1015	107.31
20	22.1108	100.87	22.0091	107.53	22.0002	107.16
Loading level of starch, phr	Peak B					
	2Θ, °	Crystallite size (L), Å	2Θ, °	Crystallite size (L), Å	2Θ, °	Crystallite size (L), Å
0	24.4151	85.42	24.3461	85.51	24.2953	80.78
5	24.3954	80.39	24.3486	83.8	24.2936	81.51
10	24.3038	75.28	24.3260	81.04	24.3317	75.90
15	24.2286	68.17	24.2748	81.58	24.3438	79.02
20	24.3397	83.34	24.2138	83.63	24.2304	82.41

The crystallite sizes of peak A and peak B for LDPE-starch blends were calculated using equation 1 and depicted in Tables 1 to 4. By referring to Table 1, the crystallite sizes of peak A and peak B for LDPE-starch blends have decreased when the starch loading level has gradually increased. This might be attributed to the addition of starch could disrupt and rupture the highly ordered arrangement of the crystallites and thus reducing the crystallite sizes of peak A and peak B. According to Table 1, when the polymer blends were immersed in 1M HCl for 3 days, the crystallite size for Peak A and Peak B were found to be higher. This indicates the realignment of LDPE chains happened when the starch was being attacked by the H^+ ions by inducing formation of new crystallites in the polymer matrix. The interaction of these newly formed crystallites with the existing crystallites could significantly increase the crystallite size.

When the immersion duration increased up to 15 days, the crystallite size of peak A and peak B for polymer blend was decreased. As the treatment duration increased, there were more radicals being formed. These radicals tend to disrupt the polymer chains and caused the breakdown of crystallites.

Table 2. The 2Θ and crystallite size of LDPE-starch blends when treated/immersed in 3M HCl solution for 3 days and 15 days

Loading level of starch, phr*	3M HCl treated/immersed duration (day)					
	0 day		3 days		15 days	
	Peak A					
	2Θ, o	Crystallite size (L), Å	2Θ, o	Crystallite size (L), Å	2Θ, o	Crystallite size (L), Å
0	22.1263	105.69	22.0422	105.51	22.0645	103.92
5	22.1323	102.93	22.1234	107.07	22.1305	104.93
10	22.0790	103.17	22.1086	108.47	22.1009	105.86
15	22.0328	92.12	22.0873	105.63	22.0441	104.92
20	22.1108	112.58	22.1321	111.07	22.0532	109.69
Loading level of starch, phr	Peak B					
	2Θ, o	Crystallite size (L), Å	2Θ, o	Crystallite size (L), Å	2Θ, o	Crystallite size (L), Å
0	24.4151	85.42	24.2868	79.82	24.2995	81.56
5	24.3954	80.39	24.3957	82.71	24.3554	77.54
10	24.3038	75.28	24.3600	82.36	24.3435	81.92
15	24.2286	68.17	24.3265	83.38	24.2666	81.47
20	24.3397	83.34	24.3664	79.68	24.2432	74.06

By referring to Table 2, the crystallite size of peak A and peak B for all LDPE-starch blends treated with 3M HCl solution showed similar trends; the crystallite size for Peak A and Peak B increased initially after 3 days, and then decreased on the 15th day. It was found that the crystallite size for peak A and peak B of plenty of polymer blends exhibited higher values compared to those degraded in 1M HCl solution. This indicates that when higher concentration of HCl solution, more H^+ ions would attack the starch in between the polymer chains [17]. This resulted more space available for the polymer chains to realign themselves into an orderly manner arrangement and formed new crystallites in polymer matrix. However, when there were more H^+ ions available, it also indicates that more free radicals ions may be taken place in the causing degradation reaction. These free radicals ions ruptured the highly ordered structures in polymer matrix and caused the reduction in crystallite size significantly. Therefore, it can be found that the crystallite size for peak A and peak B of polymer blends degraded with 3M HCl exhibited lower value than those immersed with 1M HCl.

Table 3. The 2Θ and crystallite size of LDPE-starch blends when treated/immersed in 1M NaOH solution for 3 days and 15 days

Loading level of starch, phr*	1M NaOH treated/immersed duration (day)					
	0 day		3 days		15 days	
	Peak A					
	2Θ, o	Crystallite size (L), Å	2Θ, o	Crystallite size (L), Å	2Θ, o	Crystallite size (L), Å
0	22.1263	105.69	22.1144	105.08	22.089	104.49
5	22.1323	102.93	22.0784	112.88	22.1179	104.25
10	22.079	102.53	22.0612	113.09	22.0831	107.67
15	22.0328	101.75	22.0261	117.94	22.072	112.61
20	22.1108	100.87	21.9900	118.22	22.0801	113.67
Loading level of starch, phr	Peak B					
	2Θ, o	Crystallite size (L), Å	2Θ, o	Crystallite size (L), Å	2Θ, o	Crystallite size (L), Å
0	24.4151	85.42	24.3475	80.0935	24.305	79.27
5	24.3954	80.39	24.3171	84.9108	24.391	85.89
10	24.3038	75.28	24.2981	84.3208	24.352	83.38
15	24.2286	68.17	24.2614	85.2306	24.3104	81.47
20	24.3397	83.34	24.2349	86.0800	24.3295	85.23

Table 4. The 2Θ and crystallite size of LDPE-starch blends when chemically treated in 3M NaOH solution for 3 days and 15 days

Loading level of starch, phr*	3M NaOH treated/immersed duration (day)					
	0 day		3 days		15 days	
	Peak A					
	2Θ, o	Crystallite size (L), Å	2Θ, o	Crystallite size (L), Å	2Θ, o	Crystallite size (L), Å
0	22.1263	105.69	22.0676	103.24	22.0722	110.09
5	22.1323	102.93	22.048	108.33	22.0338	108.04
10	22.0790	103.17	22.0719	113.75	22.0642	113.52
15	22.0328	92.127	22.083	115.73	22.0168	111.60
20	22.1108	112.58	22.1268	103.25	22.0784	113.12
Loading level of starch, phr	Peak B					
	2Θ, o	Crystallite size (L), Å	2Θ, o	Crystallite size (L), Å	2Θ, o	Crystallite size (L), Å
0	24.4151	85.42	24.2802	87.55	24.3312	86.05
5	24.3954	80.39	24.2679	80.76	24.2805	74.58
10	24.3038	75.28	24.3320	85.78	24.2938	85.31
15	24.2286	68.17	24.3066	88.59	24.2506	87.64
20	24.3397	83.34	24.3778	80.28	24.3156	84.46

By referring to Tables 3 and 4, the crystallite size of peak A and peak B for the LDPE-starch blends treated with 1M and 3M NaOH solution increased initially after 3 days, and then decreased on the 15th day. This indicates the reorientation of LDPE chains occurred when the starch was being attacked by the OH^- ions. Subsequently, this promotes the formation of new crystallites in the polymer matrix. The crystallite size of the polymer blends was then increased by the interaction of these newly formed crystallites with the neighbouring crystallite. When the treatment duration increased up to 15 days, the crystallite size of peak A and peak B for polymer blend was decreased. As the treatment duration increased, more free radicals have produced tremendously. These vast amounts of free radicals then disrupted the polymer chains by causing chain scissioning subsequently reduced the crystallite size.

3.3. Morphologies Observation

Figure 5 illustrates the fracture surface of LDPE-starch blends added with various loading level of starch. By referring to Figure 5(a), the discontinuity can be clearly observed on fractured surface of pristine LDPE. The formation of discontinuous phase was attributed by the applied straining stress during the tensile test. The applied straining stress could lead to the discontinuous phase of polymer matrix by tearing the polymer matrix to form thin fibrils [25]. The occurrence of long and thin fibrils indicates that the polymer is able to extent excessively before it ruptures and this shows that it has better entanglement and continuity of the structure. By referring to Figure 5(b), it was found that when 5 phr of starch was added into the polymer blends, the number of fibrils decreased greatly. This indicates the reduction of continuity of the polymer blend. On the other hand, the gaps between the matrixes can also be observed from Figure 5(c). Such observation can be related to the poor interfacial adhesion between LDPE and starch, which caused poorer tensile properties [26]. When 15 phr of starch was added into the polymer blends, the formation of short and thick fibrils can be observed on the fractured surface of polymer matrix as illustrated in Figure 5(d). The formation of fibrils in 10 phr starch added LDPE-starch matrix indicated better plastic deformation of polymer matrix but lower in rigidity. Besides that, high amount of voids can be observed on the matrix. The occurrence of voids on the matrix was mainly due to the incompatibility of LDPE and starch, resulted in a partial stress transfer between the two phases or slippage in the interfacial area between starch and LDPE matrix when the polymer blends subjected to tensile deformation [27, 28]. By referring to Figure 5(e), the polymer matrix of 20 phr starch added LDPE-starch showed greater amount of voids with bigger size. Such observations were in agreement with the tensile result discussed previously. The increment of starch composition in LDPE-starch blend reduced the tensile properties of the polymer blend. This was aligned with tensile results which deterioration occurred with increasing loading level of starch due to their incompatibility and difference in polar nature.

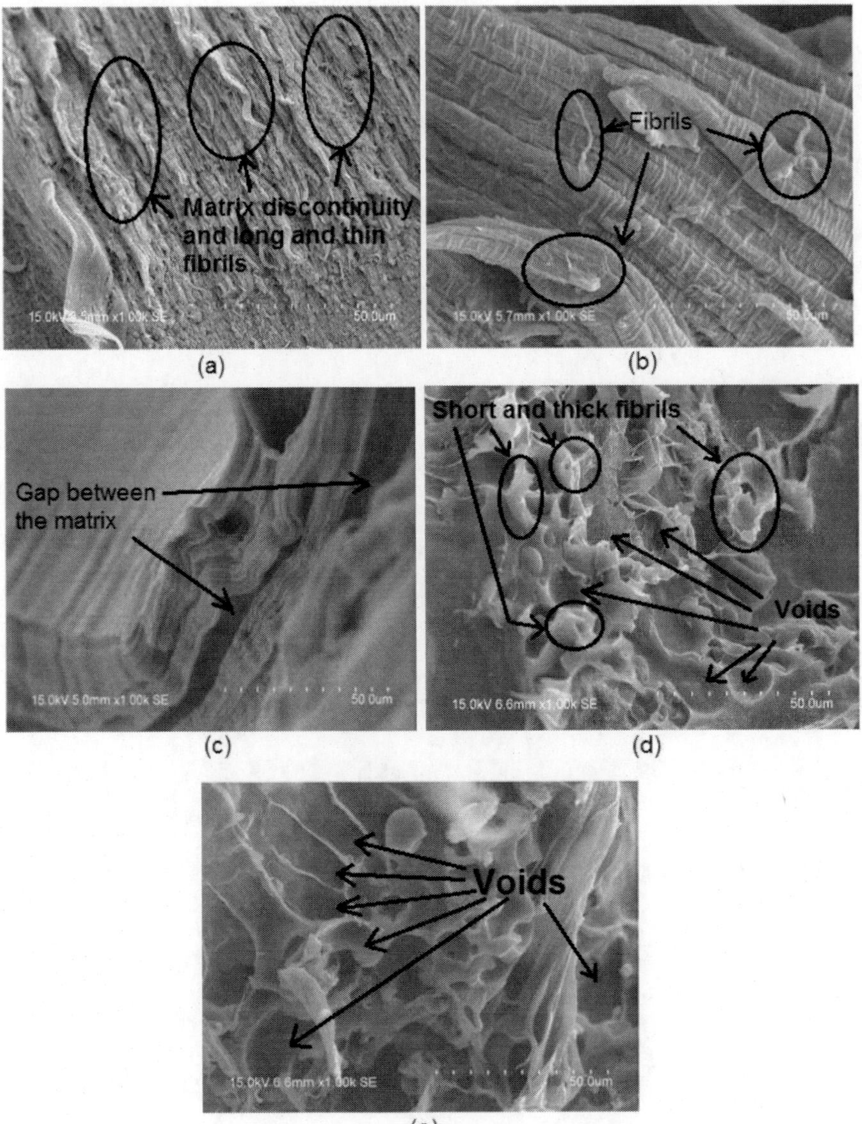

Figure 5. Morphologies observation of LDPE-starch blends added with (a) 0 phr starch, (b) 5 phr starch, (c) 10 phr starch, (d) 15 phr starch and (e) 20 phr starch under magnification of 1000 times.

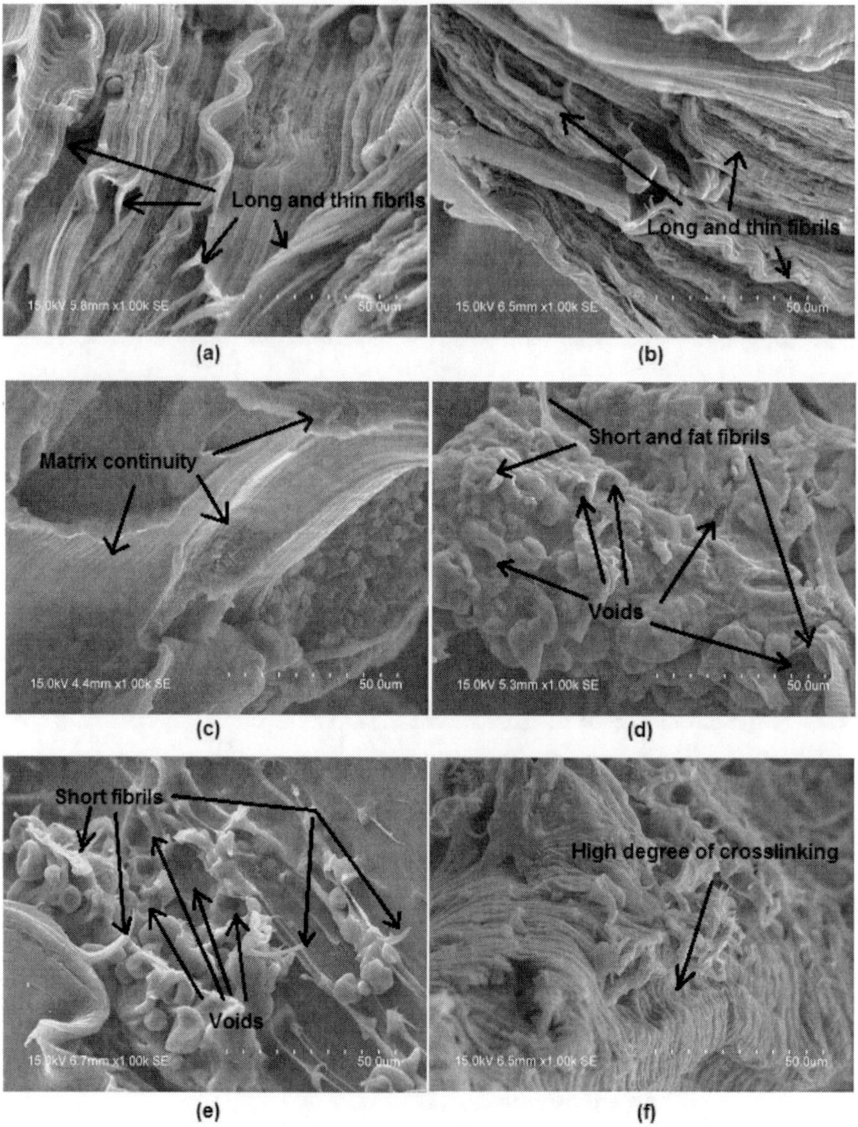

Figure 6. Morphologies observation of LDPE-starch blends added with 0 phr starch and immersed into 1M HCl for (a) 3 days, and (b) 15 days; LDPE-starch blends added with 10 phr starch and immersed into with 1M HCl for (c) 3 days, and (d) 15 days; LDPE-starch blends added with 20 phr starch and immersed into 1M HCl for (e) 3 days, and (f) 15 days under magnification of 1000×.

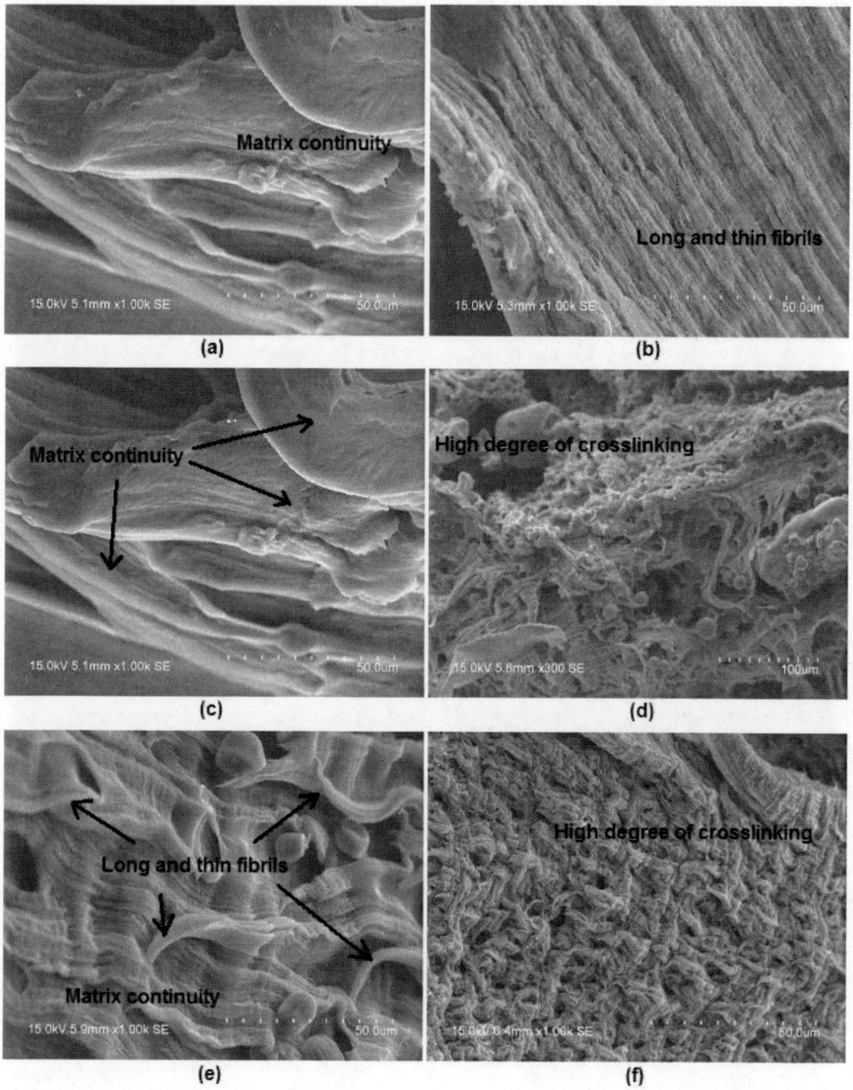

Figure 7. Morphologies observation of LDPE-starch blends added with 0 phr starch and immersed into with 3M HCl for (a) 3 days, and (b) 15 days; LDPE-starch blends added with 10 phr starch and immersed into with 3M HCl for (c) 3 days, and (d) 15 days; LDPE-starch blends added with 20 phr starch and immersed into with 3M HCl for (e) 3 days, and (f) 15 days under magnification of 1000×.

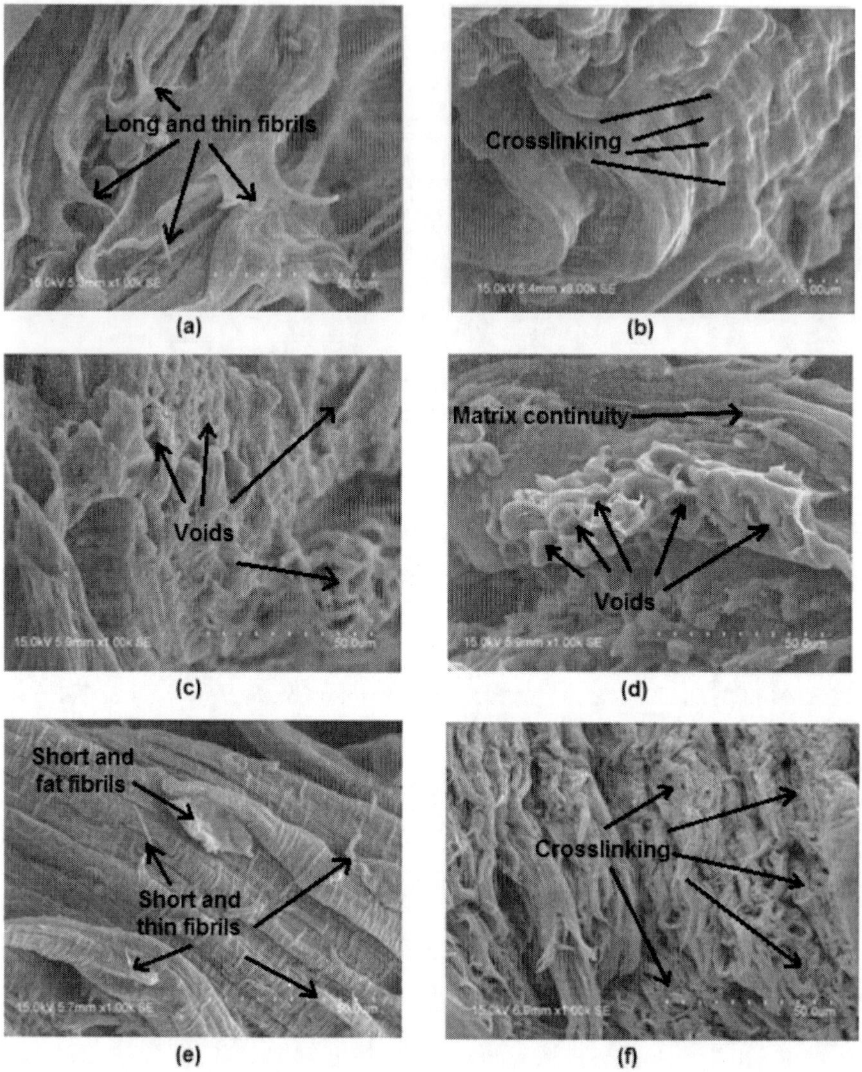

Figure 8. Morphologies observation of LDPE-starch blends added with 0 phr starch and treated/immersed with 1M NaOH for (a) 3 days, and (b) 15 days; LDPE-starch blends added with 10 phr starch and treated/immersed with 1M NaOH for (c) 3 days, and (d) 15 days; LDPE-starch blends added with 20 phr starch and treated/immersed with 1M NaOH for (e) 3 days, and (f) 15 days under magnification of 1000×.

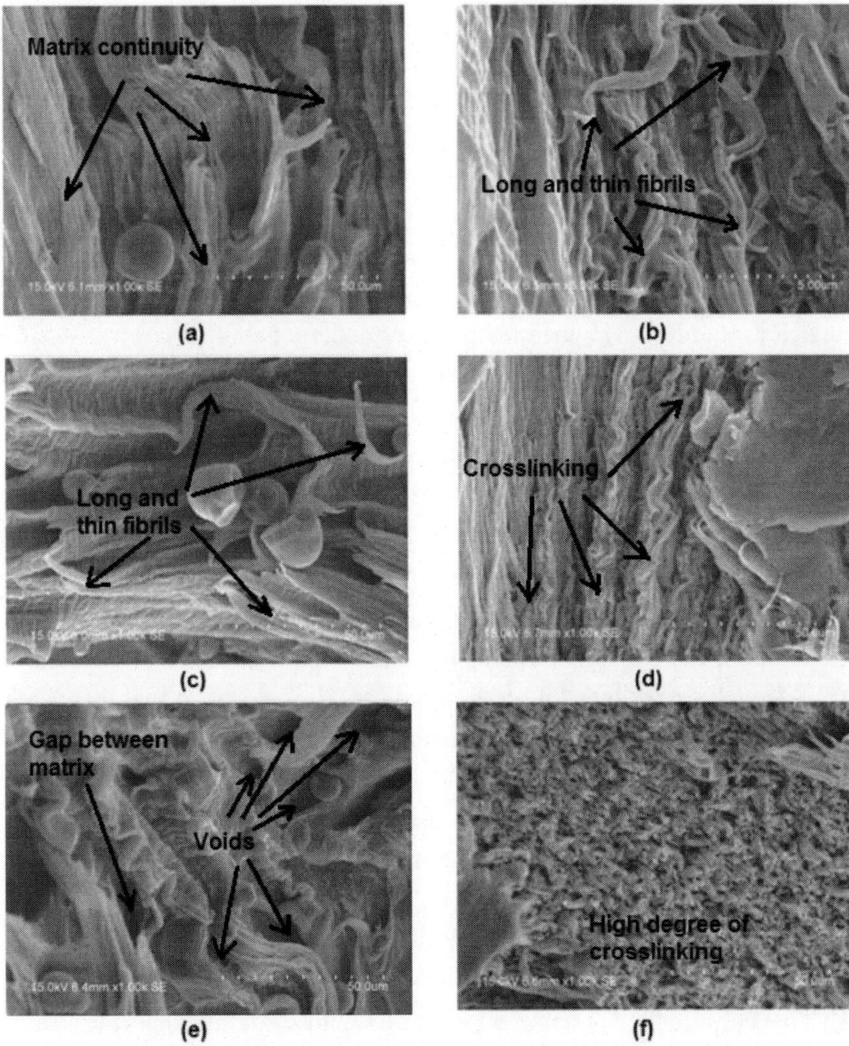

Figure 9. Morphologies observation of LDPE-starch blends added with 0 phr starch and treated/immersed with 3M NaOH for (a) 3 days, and (b) 15 days; LDPE-starch blends added with 10 phr starch and treated/immersed with 3M NaOH for (c) 3 days, and (d) 15 days; LDPE-starch blends added with 20 phr starch and treated/immersed with 3M NaOH for (e) 3 days, and (f) 15 days under magnification of 1000×.

On the other hand, the surface morphologies of all HCl acid degraded LDPE-starch blends are depicted in Figure 6 and Figure 7. By referring to Figures 6(a)-(b) and Figure 7(a)-(b), the polymer matrix of pristine LDPE was observed to be continuous with formation of long and thin fibrils when immersed for 3 days and 15 days in different concentration of HCl solution. In addition, several obvious matrix continuity in 10 phr starch added LDPE-starch matrix can be found when it was immersed for 3 days in 1M and 3M HCl (refer to Figure 6(c) and Figure 7(c)). However as depicted in Figure 6(d), when the polymer undergone treatment in 1M HCl for 15 days, the matrix showed discontinuity along with voids and short and fat fibrils [25]. This is mainly due to the attack of H^+ ions on the starch molecules, which induced formation of free radicals that attacked the LDPE matrix and resulted in poor surface morphology. On the other hand, high degree of crosslinking can be observed on the polymer matrix when the polymer blend undergone treatment for 15 days in 3M HCl (refer to Figure 7(d)). When there are greater number of H^+ ions present in the 3M HCl solution, the attacks on the starch molecules by the H^+ ions are more severe as well as produced more free radicals. It is deduced that these free radicals then undergone random chemical reaction and caused cross-linking in the polymer blend [20].

According to Figure 6(e), when the 20 phr starch added LDPE-starch blends exposed to 1M HCl solution for 3 days, short fibrils and voids can be observed on the polymer matrix. Contrarily, when the polymer blend was immersed in 3M HCl solution for 3 days, there was high matrix continuity with long and thin fibrils as illustrated in Figure 7(e). Due to high concentration of H+ ions in 3M HCl acid, they attacked the starch molecules and allowed realignment of the LDPE polymer chains in a faster manner, therefore resulted in better surface morphology. On the other hand, when the polymer blend with 20 phr starch in contact with 1M HCl solution for 15 days, high degree of crosslinking can be observed on the polymer matrix (refer to Figure 6(f)). Furthermore, Figure 7(f) shows even higher degree of crosslinking when the polymer blend exposed to 3M HCl solution for 15 days. This was mainly due to degradation induced by radical chain reactions. Such reactions caused crosslinking between the polymer chains and branching along the polymer chains [17]. When higher concentration of acid solution was being used, more H^+ ions attacked the polymer blends, which resulted more radicals being produced. Importantly, this observation was tally with the tensile results obtained.

Figure 8 and 9 illustrates the fracture surface of LDPE-starch blends immersed in NaOH solution. By referring to Figures 8(a)-(b) and 9(a)-(b), long and thin fibrils can be observed on matrix of the pristine LDPE after 3 days and 15 days immersion in different concentration of NaOH solution. Meanwhile, crosslinking can also be observed in the pristine LDPE for 15 days in 1M NaOH. Such observation indicates good tensile strength and elongation. On the other hand, voids can be observed in 10 phr starch added LDPE-starch matrix treated for 3 and 15 days in 1M NaOH as illustrated in Figure 8(c) and Figure 8(d). This is mainly due to the poor interfacial adhesion which leads to detachment of the particle/matrix connection or "pulling out" of the starch particle [27]. However, as the duration in 1M NaOH increased, the polymer blend with 10 phr starch showed matrix continuity. When the starch molecules were attacked by the OH^- ions and leave the polymer matrix, the LDPE chains were able to rearrangement themselves into a more compact structure, and thus improved the morphology of the polymer blend.

Referring to Figure 9(c), it can be observed that when the polymer blend undergone 3 days exposure in 3M NaOH, the matrix showed long and thin fibrils exhibiting good elongation. As duration prolonged, crosslinking can be observed from the polymer matrix as illustrated in Figure 9(d). From Figure 9(e), when the polymer blend with 20 phr starch undergone treatment for 3 days in 1M NaOH, short and fat fibrils can be observed on the polymer matrix. This showed poor tensile elongation of the polymer blend. In addition, crosslinking can be observed from the polymer matrix when the 20 phr starch added LDPE-starch blend immersed for 15 days in NaOH. As free radicals were being produced because of the attacks of OH^- ions on the starch molecules, there are high possibilities for crosslinking to take place [17]. On the other hand, when the polymer blend with 20 phr of starch was immersed in 3M NaOH for 3 days, voids and gaps between polymer matrixes can be observed. The lack of adhesion between LDPE and starch seen on the fracture surfaces indicated a poor interfacial interaction, which caused poorer tensile properties as illustrated in Figure 9(e). Referring to Figure 9(f), high degree of crosslinking can be observed on the polymer matrix when the 20 phr starch added polymer blend has been attacked by 3M NaOH solution for 15 days. Such observation can be related to the degradation induced by radical chain reactions which promoted crosslinking between the polymer chains and branching along the polymer chains.

CONCLUSION

The addition of starch in LDPE as the different polar elements would affect the ordered structure within the blends which resulted changes in the physico-mechanical properties of LDPE-starch blends. When the LDPE-starch blends were immersed in HCl and NaOH solutions, it can be noticed an unexpected improvement of mechanical properties occurred. This is because when starch molecules with irregular shapes and size have been dissolved/consumed, the LDPE chains became more aligned and thus improved the overall tensile strength of the polymer blend. Moreover, both XRD and SEM analyses are in agreement with the mechanical results by exhibiting higher crystallinity and visible thicken fibrils, respectively. Nonetheless, prolong exposure to acid and alkaline conditions still leading to chain scissoning that eventually causing serious degradation. This study also shows that the degradation effects are in relation to the concentration of acid and base exposure as well as the amount of starch added to the LDPE. Both reactive H+ and OH- ions tend to attack the weak structure of starch rigorously and continue to attack higher inertness of LDPE region afterwards. In conclusion, the performances of LDPE-starch blends are greatly depended on the factors of acid and alkaline condition as well as its duration of exposure. Hence, the usage of LDPE-starch blends for packaging material need to pursue in consideration of its application condition for safety of consumers.

ACKNOWLEDGMENT

The authors are grateful for the financial support from Universiti Tunku Abdul Rahman Research Fund 6200/L97.

REFERENCES

[1] Hodzic A, Shanks R. *Natural fibre composites*. Philadelphia: Woodhead Publishing; 2014.

[2] Toh WY, Rahman WAWA, Sin LT. J. *Vinyl Addit. Tech*. 2011, 17, 184-189.

[3] Cooke TF. *J. Polym. Eng*. 1990, 9, 171-212.

[4] Yeh J-T, Chai W-L, Wu C-S. *J. Polym. Eng*. 2007, 27, 483-506.

[5] Rahmat AR, Rahman WAWA, Sin LT, Yussuf AA. *Mater. Sci. Eng*. C 2009, 29, 2370-2377.

[6] Pérez E, Pérez CJ, Alvarez VA, Bernal C. *Carbohydr. Polym*. 2013. 97, 269-276.

[7] Rahman WAWA, Sin LT, Rahmat AR, Isa NM, Salleh MSN, Mokhtar M. *J. Compos. Mat*. 2011, 45, 1403-1410.

[8] Huang CY, Roan ML, Kuo MC, Lu WL. *Polym. Degrad. Stab*. 2005, 90, 95-105.

[9] Greizerstein HB, Syracuse JA, Kostynaik PJ. *Polym. Degrad. Stab*. 1993, 39, 251-259.

[10] Danjaji ID, Nawang R, Ishiaku US, Ismail H, Ishak ZAM. *Polym. Test*. 2002, 21, 75-81.

[11] Arvanitoyannis I, Biliaderis CG, Ogawa H, Kawasaki N. *Carbohydr. Polym*. 1998, 36, 89-104.

[12] Prachayawarakorn J, Sangnitidey P, Boonpasith P. *Carbohyd. Polym*. 2010, 81, 425-433.

[13] Xie F, Pollet E, Halley PJ, Avérous L. *Prog. Polym. Sci*. 2013, 38, 1590-1628.

[14] Buonocore G, Sico G, Mensitieri G. Encycl. *Food Safety* 2014, 3, 384-396.

[15] Garg S, Jana AK. *Eur. Polym. J.* 2007, 43, 3976-3987.
[16] St-Pierre N, Favis BD, Ramsay BA, Ramsay JA, Verhoogt H. *Polym.* 1997, 38, 647-655.
[17] Woishnis W, Ebnesajjad S. *Chemical Resistance of Thermoplastics*. United States: William Andrews 2012.
[18] White JR. Comptes. *Rendus. Chimie*. 2006, 9, 1396-1408.
[19] Bikiaris D, Prinos J, Panayiotou C. *Polym. Degrad. Stab*. 1996, 56, 1-9.
[20] Dintcheva NT, Jilow N, La Mantia F P. *Polym. Degrad. Stab*. 1997, 57, 191-203.
[21] Pedroso AG, Rosa DS. *Carbohydr. Polym*. 2005, 59, 1-9.
[22] Rodiguez-Gonzalez FJ, Ramsay BA, Favis BD. *Polym* 2003, 44, 1517-1726.
[23] Kumar V, Ali Y, Sonkawade RG, Dhaliwal AS. *Nuc Inst Meth Phys Res Sec B: Beam Interac Mat Atoms* 2012; 287: 10-4.
[24] Sawatari C, Kondo T. *Macromo*l. 1999, 32, 1949-1955.
[25] Bee ST, Hassan A, Ratnam CT, Tee TT, Sin LT. *Nuc. Inst. Meth. Phys. Res. Sec. B: Beam Interac. Mat. Atoms* 2013, 299, 42-50.
[26] Niska KO, Sain M. *Wood-Polymer Composites*. Philadelphia: Woodhead Publishing; 2008.
[27] Psomidaou E, Arvanitoyannis I, Biliaderis CG, Ogawa H, Kawasaki N. *Carbohydr. Polym*. 1997, 33, 227-242.
[28] Thakore I M, Desai S, Sarawade BD, Devi S. *Eur. Polym. J.* 2001, 37, 151-160.

In: Polymer-Matrix Composites ISBN: 978-1-53613-398-1
Editors: Elias Rice and Beth Sparks

Chapter 6

A NETWORK APPROACH TO THERMO-ELECTRICAL MODELING OF TEXTILE COMPOSITE STRUCTURES

Chiun-Shen Liao, Andrew Labun, Mehdi Jahandardoost and Abbas S. Milani*
School of Engineering, University of British Columbia, Kelowna, BC, Canada

ABSTRACT

In this chapter, simulation of the temperature distribution in textile composite structures is achieved by utilizing an analytical-symbolic approach. Analytical heating solutions along each yarn provide accurate solutions with far fewer nodes than numerical solutions, and require the minimum number of symbolic network nodes. The lower upper (LU) decomposition of the symbolic network scales as the cube of the number of nodes. Multiple evaluations, including iterating temperature-dependent thermal conductivity to achieve a self-consistent solution, scaled linearly with the number of nodes and hardly affected the total solution time. Results show that the memory consumption, CPU time, and solutions of

* Corresponding Author Email: abbas.milani@ubc.ca.

the proposed network calculation method, compare favorably to a finite element analysis.

Keywords: textile structures, multi-physics modeling, network calculation method, Therminator3D

1. INTRODUCTION

This study applies a new fast network-based simulation methodology to solve coupled thermo-electrical problems in textile composite structures. The solution algorithm, "Therminator3D," was originally developed to calculate the temperature distributions of interconnects due to Joule heating in microchips [1]. This algorithm demonstrates a very fast calculation on large scale Integrated Circuit layouts. To extend the benefit of this algorithm, herein we propose to use the same algorithm to the textile composite materials which their geometry is also network-based.

Composite materials are used widely in a broad range of industries to improve design properties such as lower density with higher stiffness and strength, lower effective thermal conductivities to block the heat, higher effective thermal conductivities to detect the real temperature instantly, etc. Composite materials can reach the desired mechanical properties by changing the reinforcement layouts, reducing the need for extra materials and fabrications.

In textile materials which can be knitted, woven or braided, the fiber yarns and matrix materials are consolidated during manufacturing. A number of composite components in the aerospace, automotive, marine and construction industries are built of woven textile preforms. They can give excellent mechanical properties with very low weights, but their complex weaving fabrication processes occasionally make them expensive for the particular composite manufacturers. As an alternative, non-crimp fabrics (NCFs) have proven to provide fairly good mechanical properties while maintaining low weight and providing cost effective solutions through their simplified fabrication processes. NCF composites are fabricated from preforms with multiple layers of straight fiber bundles that can be oriented in different directions and stitched together by a (warp) knitting procedure. In a study by Dransfield et al. (1998), it was reported that out-of-plane fracture toughness and damage tolerance properties of these materials are also improved through the stitching process [2].

Hind, Robitaille and Raizenne (2006) [3] conducted a numerical study on the effective thermal conductivity of a plain weave. Nordlund et al. (2006) [4] and Nordlund et al. (2004) [5] performed both numerical and experiential studies on the permeability of NCFs. Ning and Chou (1995) [6], used a closed-form representation of the transverse thermal conductivity to derive a thermal resistance network. Xu and Wirtz (2002) [7], employed a simple-to-fabricate woven mesh to describe the in-plane effective thermal conductivity of two-dimensional bonded-laminate plain-weave screen laminates. Peng, Lu and Balendra (2004) [8] studied the FE simulation of the blanking of electrically heated engineering materials. They also used thermo-electrical simulation with FEA implementation.

For multi-axial textile material simulation, Zhao and Hoa (2003) [9] performed thermal deformation behavior of triaxial woven fabric composites using super-finite-elements. They developed a super-element which is one kind of unit-cell to represent the model of simulation. They assumed the thermal expansion coefficients remain constant within a certain region of the panel in order to reach constrain conditions. Results indicate that this super-model is good for specific simulations.

Xiong, Shenoi and Gao (2009) [10] proposed an analytical model to predict residual thermal stress in two-dimensional plain weave composites. They also applied ideal representative volume elements to represent their plain weave fabric. There is no advanced proposal to present the capacity of this model in large scale weave composites.

The majority of the numerical studies on the mechanical and/or thermal analysis of fabrics have been based on the meso-level FEA analysis of unit cells representing repetitive (network) patterns in the fabrics. In practice, however, fabric unit cells are not perfectly repetitive, due to uncertainties during fabric fabrication processes and/or unsymmetrical loading conditions for a given problem. They may require performing analyses on the larger representative volumes. As the size of the problem scales, the solution complexity and the simulation run time can pose a problem. In addition, the selection of the position of a unit cell and the associated boundary conditions can become highly important.

The layout of NCF materials is similar to that of printed circuit boards (PCB). The architecture is layer by layer and yarn-crossed fabrication. For the woven materials, the layout is not identical but it is similar to the printed circuit boards. The woven structure can be interpreted into the same layouts of interconnects with suitable twisted (angular) coupling values. The simulation of full-layout textile material properties in commercial CAD/FEM tools including the layout drawing, CPU time, and memories consumes several resources. Li et al. (2008) [11] conducted a research study to use the concept of Finite shell model of ABAQUS to simulate the bending behavior in Ultra-thin and Multi-layer PCB using woven fiber composite. They concluded that the bending stiffness of the multi-layer PCBs can be enhanced by the high fiber bundles of the woven fiber in the fill direction. However, they did not discuss the consumed computation resources of the simulations.

In this study, a new method was proposed to employ "Therminator3D" for textile materials in order to reach more rapid multi-physics simulations with low CPU time and memory. Therminator3D is capable of simulating temperature distributions of the entire thermal network rapidly and accurately, without relying on the simplified unit cell descriptions. Therminator3D applies network simulation concepts with technological parameters extracted from CAD layout descriptions to the entire network (similar to FE models).

2. Proposed Methodology

We propose a network-based simulation, Therminator3D, which is an approach based on the information of the whole network. It has a concept of network calculation on nodes and corresponding connections of the entire network. Each node has the detail information between adjacent nodes and itself. To interpret the textile material simulations, there are netlist files which describe the coordinates of components, the values of capacitance and resistance, and the properties of materials. The list of a net provides detailed data including the values of resistors, properties of the materials and the information of the adjacent nodes. The net-based simulation can be composed of several nets and those nets may be coupled. At the net-based approach, each net is solved individually then the nets are assembled from the layout database to form the entire network. Therefore, by using the net-based approach, the total calculation time does not increase with the entire model size, but it rather increases with the size of each net and the number of nets.

There are some assumptions which are applied to the applications of this study:

- This research only calculated heat conduction behavior, therefore; heat convection and heat radiation are not considered in this study.
- In this study, only simulated steady-state problems have been considered.
- The temperature in each conductor varies only in the lengthwise direction.
- The property of via represents the degree of contact between the column and row yarns. For the applications presented herein, it was assumed that those contacts are perfectly connected with columns and rows.

2.1. The Analytical Solution of Networked Elements

For the analysis of the textile material, the solutions of the one-dimensional heat equation are derived on a rectangular fiber "thermal resistor" segment and on vias which represent the contact of overlapping fibers, Figure 1. The networks that are formed by these resistors and vias, layout is shown in Figure 2, are treated in the same way as the electrical resistors that are used in metal interconnect Electronmigration reliability

(EMR) analysis [12]. Each segment has a resistance R_n which is given by the following equation:

$$R_n = \frac{L_n}{k_n A_n} \tag{1}$$

Where L is its length (m), k is its conductivity ($\Omega - m$), and A is its cross-sectional area (m^2).

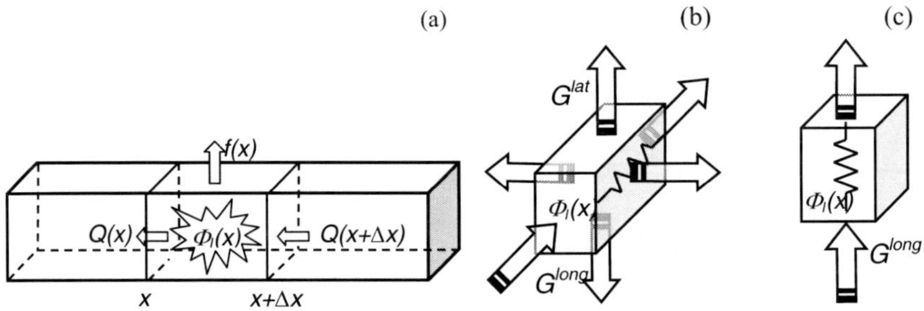

Figure 1. Heat conservation model and capacitances. (a) Heat conservation within a differential slice of a one-dimensional thermal resistor accounts for longitudinal flux, lateral flux, and Joule self-heat. (b) Horizontal resistors have G^{lat} and G^{long}, while (c) via resistors only have G^{lat}.

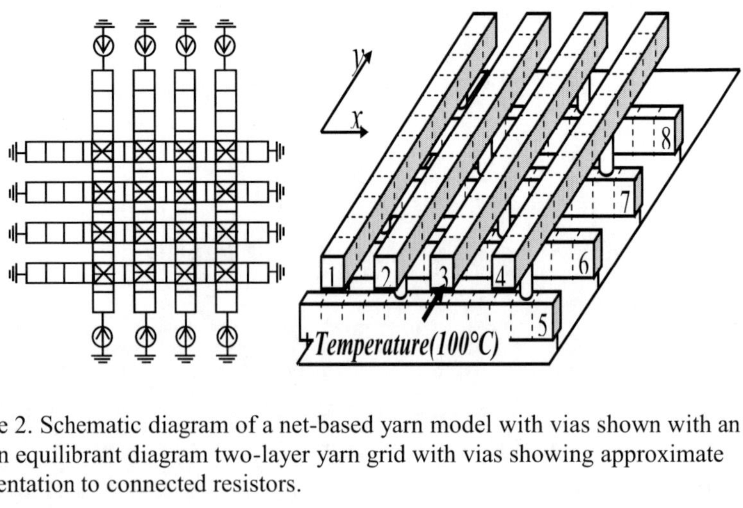

Figure 2. Schematic diagram of a net-based yarn model with vias shown with an 'X' and an equilibrant diagram two-layer yarn grid with vias showing approximate segmentation to connected resistors.

Consider a segment of length Δx on the thermal resistor network as shown in Figure 1. Its heat conservation can be derived as equation (2) [1]:

$$mC_p \frac{\partial T}{\partial t} = Q(x + D_x) - Q(x) + F_t(x)D_x - f(x)D_x \tag{2}$$

The longitudinal heat transfer rate, *Q*, is opposite to the increasing direction-x; G^{long} is a longitudinal thermal conductance derived by integrating thermal conductivity k^{long} over the resistor cross-section. The lateral heat rate per unit length (*W/m*) is written as $f(x) = G^{lat}(T(x)-T^{ref})$ with the lateral thermal conductance G^{lat} representing the diffusion of heat flux from the sides of the resistor with unit length, through a dielectric with thermal conductivity k^{lat}, to a uniform reference temperature T^{ref}. Parasitic interconnect capacitance *C* is analogous to G^{lat} and both may be calculated using the same technique. The segment mass is *m* and *Cp* denotes the specific heat. The Joule self-heat per unit length Φ_l (W/m) generated by a current in the circuit is $\Phi_l = I_{rms}^2 R_l$. I_{rms} is the root mean square of the current.

In a steady-state case, *T* (*x*, *t*) is a function of x only, and when $\Delta x \rightarrow 0$, Equation (2) becomes a second order ordinary differential equation (ODE):

$$\Phi_l = -G^{long}\frac{d^2T}{dx^2} + G^{lat}(T - T^{ref}) \tag{3}$$

The solution on each resistor segment with the boundary conditions of $T(x = 0) = T_0$ and $T(x = L) = T_L$, is then given by [1];

$$T(x) = \left(\frac{T_L - T^{\infty} - (T_o - T^{\infty})\cosh(\xi L)}{\sinh(\xi L)}\right)\sinh(\xi x) + (T_o - T^{\infty})\cosh(\xi x) + T^{\infty} \tag{4}$$

Where $\xi = \surd(G^{lat}/G^{long})$ and T^{∞} is a constant term,

$$T^{\infty} = \frac{F_l}{G^{lat}} + T^{ref} \tag{5}$$

For specific resistors such as vias for which $G^{lat} = 0$, the solution becomes

$$T(x) = -\frac{I_{rms}^2 R x^2}{2G^{long}} + \left(\frac{T_L - T_o}{L} + \frac{1}{2} F_l \frac{L}{G^{long}}\right) x + T_o \tag{6}$$

Equation (4) and (6) have been applied to estimation the thermal distribution of the multilevel interconnects [13]. In the terminator 3D, electrical and thermal resistances are taken as constants within a resistor given a temperature. However, each resistor has its own values, depending on its average temperature.

The network numerical solutions are iterated over temperature and current until a self-consistent state (equilibrium) is achieved.

The textile fiber model can be interpreted as an equivalent interconnect RC model, Figure 3. The network nodes correspond to the resistor endpoints. The system of analytical boundary value equations (5) and (6) can be evaluated over the same RC network by assigning the appropriate values to the circuit symbols in Figure 3. From equations (5) and (6), for yarn resistors, the equivalent quantities are:

$$\alpha = G^{long} \xi / \sinh \ \xi L \tag{7}$$

$$\Psi = 2\Phi_l \ \cosh \ \xi L \ -1 \ / \ \xi \sinh \ \xi L \tag{8}$$

$$\beta = \alpha \ \cosh \ \xi L \ -1 \tag{9}$$

For vias:

$$\alpha = G^{long} \ / \ L \tag{10}$$

$$\Psi = \Phi_l \ / \ L \tag{11}$$

Connected resistors have common temperatures at their junctions, and the heat flux along their length is conserved. Analytical trajectories of equation (5) have been previously applied to study thermal scaling of the interconnect (yarn) architectures [14].

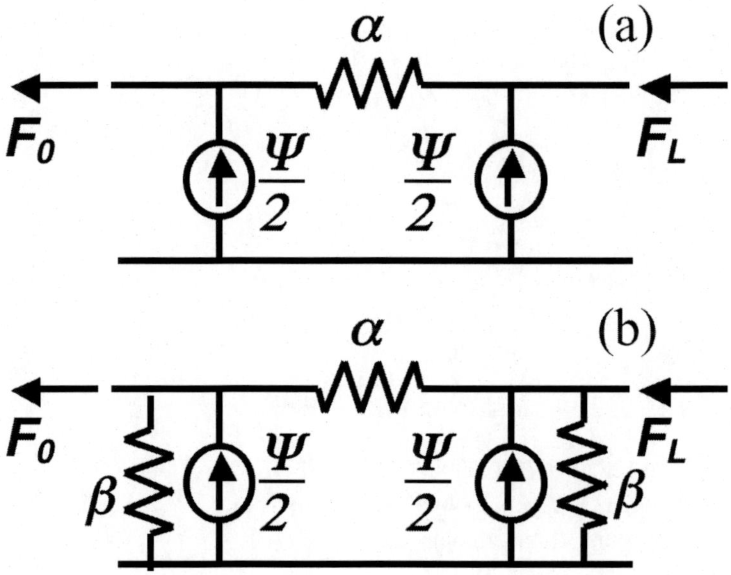

Figure 3. EMR methodology pi network diagrams. Each of the resistors that form a net is represented by the equivalent pi network. Network (a) is used for via resistors and network (b) for other resistors, but with conductance $\beta = 0$ in electrical analysis and $\beta > 0$ in thermal analysis or could represent any other mechanical property.

2.1.1. The Format of Input Files

The two input files for the Therminator3D are the netlist file and the coupling file. The formats of these two files are shown in Table 1. The netlist file is formed by the resistor and node numbers, the coordinates, length, and thermal resistance values. Every line in the coupling file consists "CAP," the resistor number, the "coupled to" resistor number, capacitance value, and the value of lateral thermal conductance G^{lat}.

Table 1. Therminator3D input file format and typical values

netlist file (resistor node0 nodeL R # xlowerleft ylowerleft xupperright yupperright L R^{long}) (current source node0 nodeL *Irms*)
. . . ... RES27 25 26 83.046936 # 40 -20 40 -10 10.000000 3433.110212 RES28 26 27 83.046936 # 40 -10 40 0 10.000000 3433.110212 . . . . . . AMP1 0 23 0.000000 RESAMP1 0 23 3.000000 # 35 -45 45 -35 1.000000 0.0001 . . . . RES4-27 4 27 3.000000 # 35 -5 45 5 1.000000 0.001 ......
coupling file (CAP RESISTOR {RESISTOR|0(SUBSTRATE)}*c* # G^{lat}
...... CAP RES27 0 12.666667 # 0.00000040086361458 CAP RES28 0 6.333333 # 0.00000020043180686 CAP RES28 RES4 9.500000 # 0.00000030064771072 CAP RES28 RES5 9.500000 # 0.00000030064771072 ......

The coupling file presents the heat conductance to the neighboring dielectric, substrate, and resistors. The pair-wise G^{lat} through the dielectric between each resistor and its neighboring resistors (analogous to their coupling capacitance) can be extracted from the layout by a parasitic capacitance extractor and summed to form the total G^{lat} for that resistor, as shown in Figure 4. The equations for $K_{in-plane}$ and $K_{cross-plane}$ are [15]:

$$k_{in-plane} = \frac{1}{(1-x)(k_v P + k_f(1-P))^{-1} + \frac{P_x}{k_v} + \frac{(1-P)x}{k_f}} \tag{12}$$

$$k_{cross-plane} = (1-x)(\frac{P}{k_v}+\frac{P}{k_f})^{-1} + (k_v - k_f)Px + k_f x \quad (13)$$

where P is the porosity of the matrix, k_v is the thermal conductivity of the inclusion medium, k_f is the thermal conductivity of the host medium, *and r* is the ratio of the parallel component area to the total area.

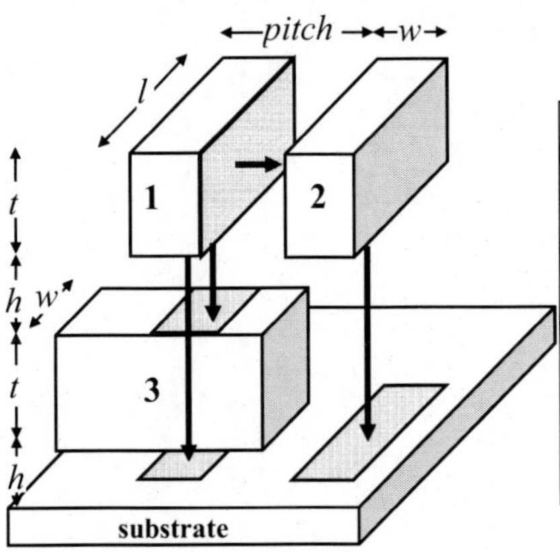

$$G_1^{lat} l T^{ref} = G_{1,2}^{lat} l T_2^{ave} + G_{1,3}^{lat} w T_3^{ave} + G_{1,substrate}^{lat} (l-w) T_{substrate}$$

$$G_{1,2}^{lat} = K_{in-plane} \frac{t}{pitch - w}$$

$$G_{1,3}^{lat} = K_{cross-plane} \frac{w}{h}$$

$$G_{1,substrate}^{lat} = K_{cross-plane} \frac{w}{2h+t}$$

Figure 4. Typical Lateral thermal conductance model in IC Interconnects. Lateral thermal conductance G^{lat} between each resistor and its assumed uniform background temperature T^{ref} is the weighted average of temperatures to neighboring resistors and the substrate, using simple area models of conductance. Anisotropic thermal conductance is indicated by $K_{in-plane}$ and $K_{cross-plane}$ values [21].

Although a simple area model is used in this study, the more sophisticated coupling models are found in most capacitance extractors which could potentially be used for more accurate results if supplied with appropriate thermal conductivity parameters.

For example, in application to woven materials (Figure 5), the simple area model is not appropriate to describe G^{lat} to represent the phenomenon. Currently, there are no experimental data to achieve the formulization of Glat as related to the woven structures and coupling areas, and the matrix between the fibers. In this research, we initially estimate Glat by the simple area model in Figure 5 then calibrate the values of G^{lat} with the result of numerical simulation using ABAQUS. The G^{lat} for the twist part with angular section coupled to yarn 3, G^{lat}_{twist}, is around 0.15 of $K_{cross-plane} w / 2h$. Then the result of Therminator3D simulation is closest to the result of ABAQUS simulation. That is because the angular section enhances the coupling effects to adjacent yarns as the distant between yarns changes.

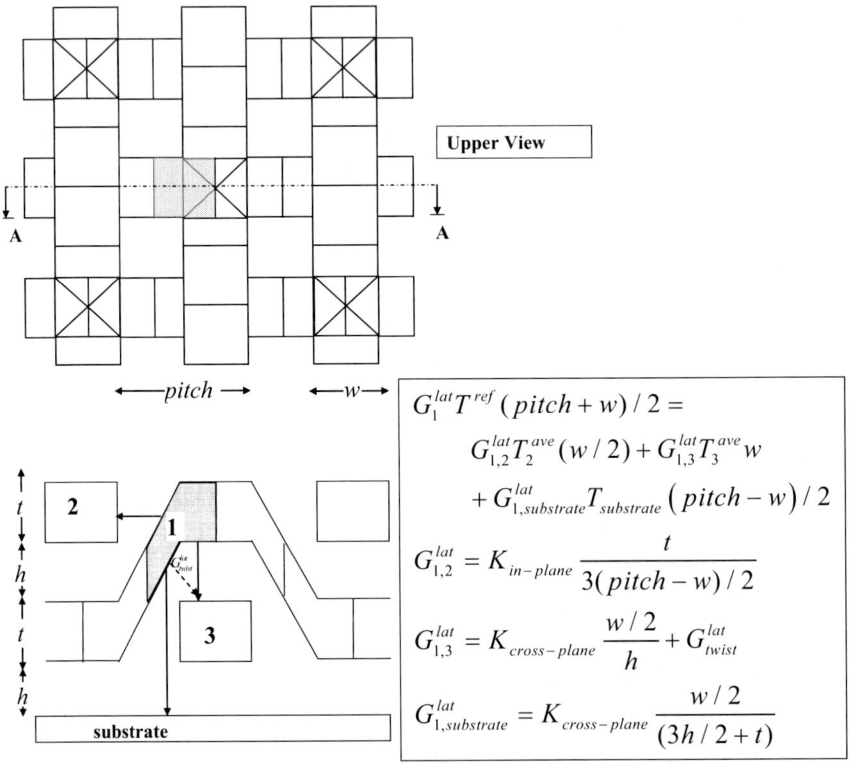

$$G^{lat}_{1} T^{ref} (pitch + w) / 2 = G^{lat}_{1,2} T^{ave}_{2} (w / 2) + G^{lat}_{1,3} T^{ave}_{3} w + G^{lat}_{1,substrate} T_{substrate} \left(pitch - w\right) / 2$$

$$G^{lat}_{1,2} = K_{in-plane} \frac{t}{3(pitch - w) / 2}$$

$$G^{lat}_{1,3} = K_{cross-plane} \frac{w / 2}{h} + G^{lat}_{twist}$$

$$G^{lat}_{1,substrate} = K_{cross-plane} \frac{w / 2}{(3h / 2 + t)}$$

Figure 5. Unit-Cell lateral thermal conductance for woven case (shown by the capacitor symbols).

2.2. Symbolic Network Analysis Solution

The sparse *N x N* system of linear algebraic equations can be expressed by the following equation:

$$Az = b \tag{14}$$

It can be decomposed to lower and upper triangular matrices L and U with order $O(N^3)$ operations by the Crout algorithm. The system can then be evaluated for z by two consecutive back-substitution operations ($O(N)$ if null operations are not performed):

$$Az = (LU)z = L(Uz) = b \tag{15}$$

$$Ly = b \tag{16}$$

$$Uz = y \tag{17}$$

The elements l_{ij} and u_{ij} of L and U can be found by equations (11), (12), and (13) over the elements a_{ij} of A [16, 17].

$$\begin{aligned} &I_{ij} = 0 \; for \; i < j, u_{ij} = 1 \; for \; 1 \le i \le N \\ &and \; u_{ij} = 0 \; for \; i > j, where \; 1 \le i, j \le N \end{aligned} \tag{18}$$

$$l_{im} = a_{im} - \sum_{\mu=1}^{m-1} l_{i\mu} u_{\mu m}, i = m, m+1, \ldots, N \tag{19}$$

$$u_{mj} = \frac{1}{l_{mm}} (a_{mj} - \sum_{\mu=1}^{m-1} l_{m\mu} u_{\mu j}), j = m+1, m+2, \ldots, N \tag{20}$$

Therminator3D, a C language program, applies dynamic storage allocation and data structures to store the symbolic structure of the solution. Each resistor in the net-list file is given a data structure containing all numerical data and results for the problem to be solved. All the resistor structures are referenced through 'hash' tables. The problem matrix A is represented by a linked list of $O(N)$ non-null elements of A. Figure 6 shows the distribution of elements in A. The cross marks represent the non-zero values. The distribution of Figure 6 is the least-sparse case with all vias between columns and rows. The vector b consists of links to heat source elements and boundary conditions. Matrix A is decomposed to the L and U linked lists by the Crout factorization formulas. Vectors z and y are constructed as linked lists pointing to the components of L, U, and b. After the entire system of linked lists has been created, it is straightforward to evaluate the specific physical problem (mechanical, thermal, electrical, etc.). For thermal problems, the temperature trajectories $T(x)$ are determined after the z vector (the temperature of each node) has been found. Average and maximum values are stored in the data structure for convenience. The $O(N)$ evaluation can be iterated for nonlinear problems (for example, when k^{long} is temperature-dependent) or for multiphysics problems.

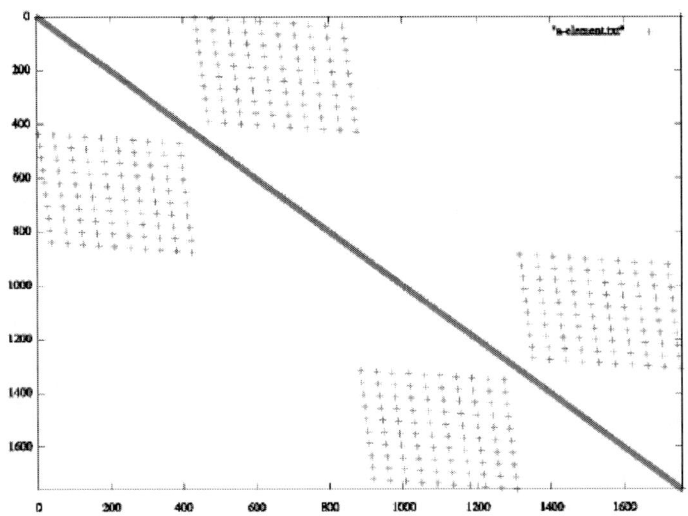

Figure 6. The distribution of problem matrix A (the 20 columns and 20 rows case with via between each row and column).

Therminator3D applies the Crout algorithm for LU decomposition and back-substitution to the system and records only the non-null operations, therefore; it leads to forming a generic symbolic solution for the entire net. This symbolic solution is evaluated with electrical parameters to solve the electrical problem first, and is then evaluated again with thermal parameters to solve the resulting Joule heat problem. The resistance of each resistor can increase with increasing the temperature. The final resistance and temperature are therefore found iteratively. Also, thermal conductance through the insulator between adjacent resistors on different nets may be extracted in the same way as capacitive coupling for IC interconnect. By composing the ambient, reference temperature for each resistor from temperatures of resistors to which it couples through the dielectric, it is possible to iteratively couple separate nets' temperature trajectories. To form the symbolic solution requires $O(N^3)$ operations, but for a sparse system such as an interconnect network there are only $O(N)$ non-null operations (where N is the number of nodes). Thus, the electrical and thermal evaluations require only $O(N)$ operations. Therminator3D requires a database that contains network information, such as the nodal relationships, coupling between resistors, input sources, and boundary conditions. A netlist file extracted from IC layout with a standard CAD tool would be the most common source of this information. In Therminator3D, the symbolic solution procedure takes most of the CPU time to allocate the dynamic memory in order to store the symbolic structure of the A, L and U matrices. After formed symbolic solution, then the program starts the numerical solution loop until the convergence of coupling electrical and thermal calculations are achieved.

2.3. Numerical Solution

The analytical solution procedure involves the coupling of thermal and electrical calculation until the convergence of temperature solution for the steady-state condition is reached. After the symbolic solution is done, the L and U matrices in linked-list forms pass to the analytical solution loop.

For the electrical problem, the pi equivalent network values α, β, and Ψ are considered from electrical circuit values for each resistor as described in Figure 3. For our applications, the electrical conductivity of resin material is almost zero, therefore; $\beta = 0$. The value of non-zero a_{ij} is the sum of the α field of linked resistors. The elements of y matrix and z matrix are calculated by the back-substitution.

For the thermal problem, the pi equivalent network values α, β, and Ψ are considered from thermal and electrical current values for each resistor using equations **(7)**, **(8)**, and **(9)**, and the evaluation procedure is repeated. The z array (nodal temperature in the thermal problem) is then found. The temperature of resistors is updated and coupled to form the T^{ref} (uniform thermal reference):

$$G_{n_m}^{lat}(T_{n_m} - T_{n_m}^{ref}) = \sum_{\nu=0}^{N}\sum_{\mu=1}^{M_\nu} G_{n_m\nu_\mu}^{lat}(T_{n_m} - \overline{T_{\nu_\mu}}) \tag{21}$$

$$T_{n_m}^{ref} = \frac{\sum_{\nu=0}^{N}\sum_{\mu=1}^{M_\nu} G_{n_m\nu_\mu}^{lat}\overline{T_{\nu_\mu}}}{G_{n_m}^{lat}} \tag{22}$$

where

$$\overline{T} = \frac{1}{L}\int_{x=0}^{L} T(x)dx, \quad \textit{for each resistor} \tag{23}$$

Since T^{ef} is also updated, the temperatures of nodes of the entire net should be recalculated through electrical and thermal solutions until convergence is reached.

Figure 7 shows the flowchart of the Therminator3D method. The therminator3D program uses one package to calculate both electrical and thermal problems using the same symbolic solution procedure. Therminator3D analytical approach may be more suitable for large systems by making only minor modifications from the EMR verification methodology [12, 18]. The modified RC-network model describes all possible properties of a resistor or component. Additional coupling data represent that the behavior of thermal coupling could also represent other mechanical coupling properties. This gives the Therminator3D method a potential to solve large networks of multi-physics problems with high speed and accuracy.

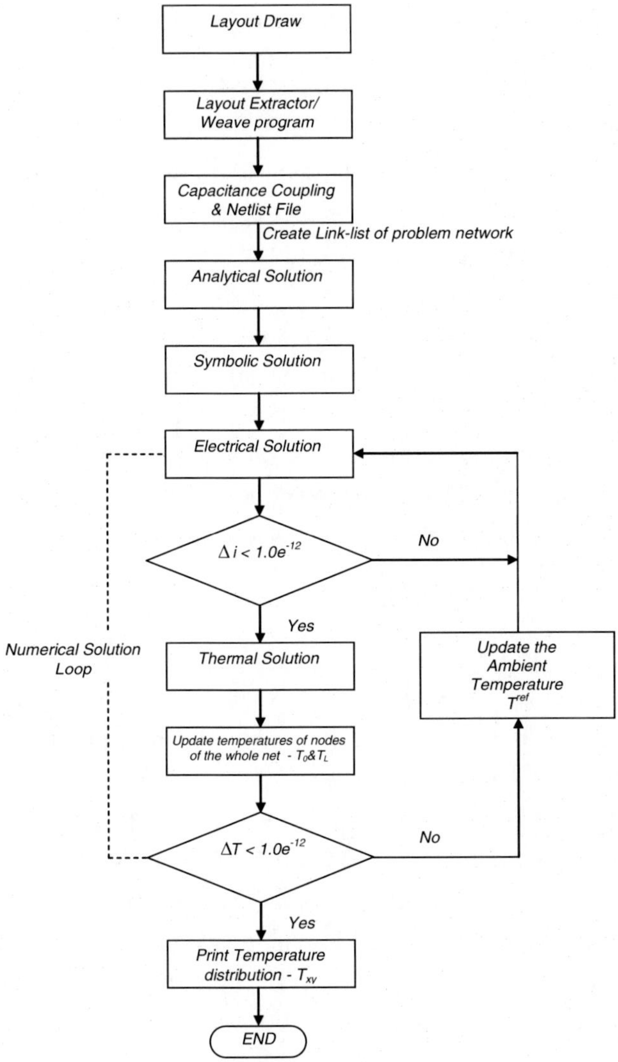

Figure 7. Flowchart of Therminator3D simulation.

Therminator3D uses the fixed-point iteration algorithm to evaluate the numerical solution. The proof of convergence of Therminator3D method is described in Liao et al. 2010 [19].

2.4. Simulation Sequence of Therminator3D

The sequence of simulation in the Therminator3D method is starting from the layout drawing. There are many free open-source software which could be used for this purpose such as Magic, Technology Computer-aid Design software in VLSI field. The second step is extracting the detailed information of netlist and capacitance of the whole network. The following step is to solve the analytical solution. From the analytical solution, this program then will form the symbolical solution with problem matrix A and the L and U matrices. After the symbolic solution is done, Therminator3D program starts the procedure of numerical solutions to solve the entire coupling electrical and thermal by achieving the convergence of both calculations. For detail information about the convergence of the Therminator3D method, please refer to Liao et al. 2010 [19].

3. Applications and Results

The results of thermo-electrical simulations from the Therminator3D method will be compared with those of ABAQUS simulations in different examples. For all simulation examples, we have used an identical MacBook machine, which has MAC OSX 10.5 OS, 4G RAM, and Intel Core 2 Duo 2.4 GHz. The compiler for the Therminator3D program is gcc version 4.3. ABAQUS is installed in windows XP pro system in the same MacBook. There are example cases of non-crimp fabric composites, woven fabric composites and 20 by 20 large network with temperature boundary conditions and Joule self-heating current. The material properties used in this study are shown in Table 2. The dimensions of each yarn and vias for different cases are shown in Table 3.

In ABAQUS, we use the steady-state coupled thermal-electric simulation with automatic incrementation. The method of matrix storage for equation solver is unsymmetrical. While executes the coupled thermal-electrical analysis, the ABAQUS will trigger the unsymmetrical matrix storage and solution scheme. We use the default setup, propagate from previous step, while convert sever discontinuity iterations. The setup of default load variation with time is "Ramp linearly over step." Extrapolation of previous state at start of each increment is using linear extrapolation.

Table 2. Material properties used in simulations

	Material	Thermal Conductivity (W/m-K)	Electrical conductivity (ohm-µm)
Yarn (Row and Column)	Bulk-Cu	401	22
Vias	Bulk-Cu	401	3
Resin (Matrix)	SiO2	1.4	10^{-14}

Table 3. The dimension of yarns

	Height (µm)	Width (µm)	Length (µm)
NCF	10	10	100
Woven	10	10	180
Large Scale Network (20x20 nets)	10	10	420
Via	10	10	10

The mesh method and the element type of yarns are linear hexahedron and type DC3D8E. The mesh method and the element type of resin are linear tetrahedron and type DC3D4E.

The nonlinear solution technique which we used for coupled thermal-electrical analysis in ABAQUS is the full Newton method. From the ABAQUS Theory Manual [20], for N nonlinear system equations:

$$F^N(u^M) = 0 \tag{24}$$

where F^N is the component with N^{th} variable in the problem and u^M is the value of Mth variable. The problem is to solve:

$$K_i^{NP} c_{i+1}^P = -F_i^N \tag{25}$$

where

$$K_i^{NP} = \frac{\partial F^N}{\partial u^P}\left(u_i^M\right) \tag{26}$$

is the Jacobian matrix and,

$$F_i^N = F^N\left(u_i^M\right) \tag{27}$$

the next approximation to the solution is:

$$u_{i+1}^M = u_i^M + c_{i+1}^M \tag{28}$$

and the iteration continues until whole system achieves by the quadratic convergence of the full Newton method.

3.1. Non-Crimp Fabric Network

The overall layout of non-crimp fabric is made by the layers with stitches that fix the fabric shape. We can conveniently translate the whole non-crimp fabric network into an interconnect network with original techniques used in Integrated Circuit field. The yarn of NCF materials is analogous to the interconnect of IC circuits. The vias, which can be assigned different values in Therminator3D, express the contact behavior between yarns in NCF. The dielectric in IC circuits is analogous to the resin material in NCF networks. In this research, overlapping yarns are assumed to be in perfect contact. The resin material property is analogous to the 0% porosity SiO_2.

An example was chosen to demonstrate the application of T3D for the thermo-electrical analysis of fabrics (e.g., for E-textiles where electronic components are embedded into a fabric structure). The NCF geometry of four-by-four, two-layer grid was considered and an electric current (0.001 Amp) was applied to yarn 7 (Figure 8). Each yarn is 10 μm x 10 μm x 180 μm. In this case, the Joule heat creates a body heat flux through out yarn 7. The result of the network calculation is shown in Figure 9 and the result of ABAQUS simulation is shown in Figure 10. Self-consistent thermo-electrical calculation with temperature-dependent thermal and electrical resistance required only 0.096 sec CPU time for the solver to fully converge. Figure 11 shows the temperature comparison on all 8 yarns. The temperature distribution lines of ABAQUS and Therminator3D along all 8 yarns are reasonably matched with each other. The comparison of CPU time and memory required by Therminator3D and ABAQUS is shown in Table 4.

Table 4. Comparison of CPU time required by Therminator3D and ABAQUS

		Vias	No. of Elements	CPU Time(sec)
NCF	T3D	yes	96 + 8 (Vias) (104 nodes)	0.096
	ABAQUS	yes	11574 (6404 nodes)	13.1
Woven	T3D	yes	96 + 8 (Vias) +2 (Sky) (106 nodes)	0.06
	ABAQUS	yes	12893 (3751 nodes)	8.1
Large Scale NCF	T3D	yes	1680 + 400 (Vias) (1680 nodes)	181.19 (179 seconds for symbolic solution, 2 seconds for numerical solution)
	ABAQUS	yes	52263 (45538nodes)	391.70
	T3D	no	1680 (1680 nodes)	181.18
	ABAQUS	no	52263 (45538 nodes)	325.40

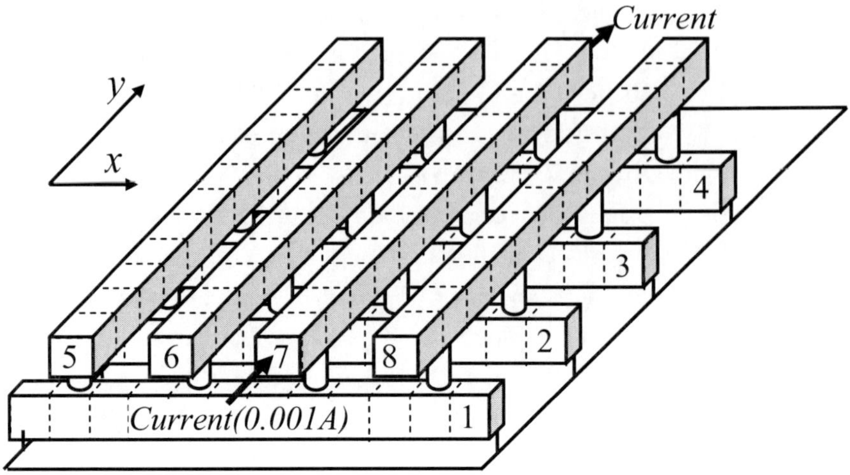

Figure 8. A two-layer textile grid layout with vias and Joule heat generated on yarn 7.

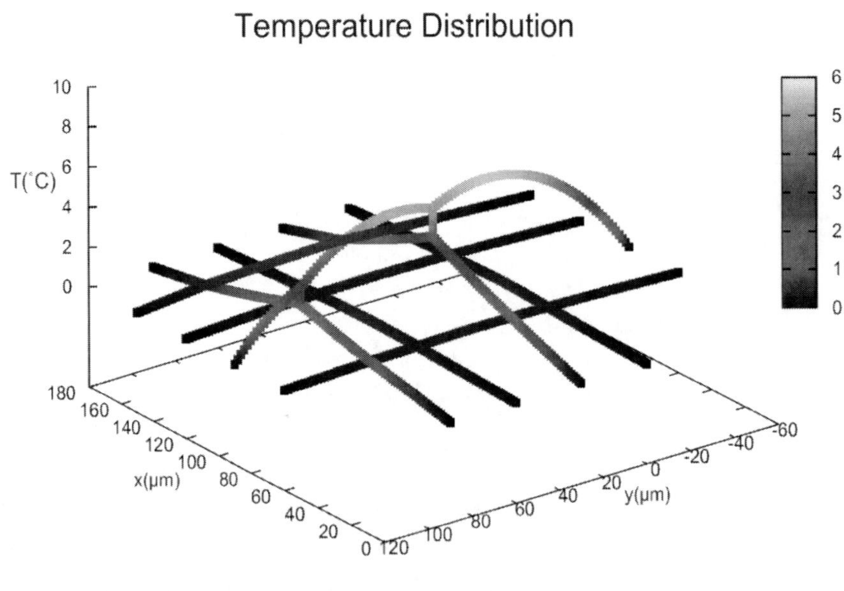

Figure 9. T3D implementation of NCF layout in Figure 8.

Figure 10. ABAQUS implementation of NCF layout in Figure 8.

3.2. Woven Fabric Network

Woven fabric composite materials are composed of interlacing yarns with good mechanical properties, especially for impact applications. The exact shape of the woven fibers is rather difficult to reproduce using CAD tools. Descriptions of the exact contact behaviors between fibers are also challenging. In Therminator3D however, the shape of yarns can be represented by switching the coordinates of nodes in the NCF case and replacing new coupling data for the thermal and electrical capacitances. In this section, we have made a comparison between the results of the two simulation (ABAQUS and Therminator3D) to show the potential of Therminater3D for thermal and electrical evaluations of woven materials.

The layout of the fabric considered is four by four dry plain weave yarns with partial vias shown in Figure 12. The dimension of each yarn is 10μmx10μmx180μm (WxDxL). The material properties used in this study are listed in Table 5.2. A current (0.001 AMP) was applied to yarn 5. The resulting temperature distributions of T3D and ABAQUS are shown in Figure 13 and Figure 14, respectively.

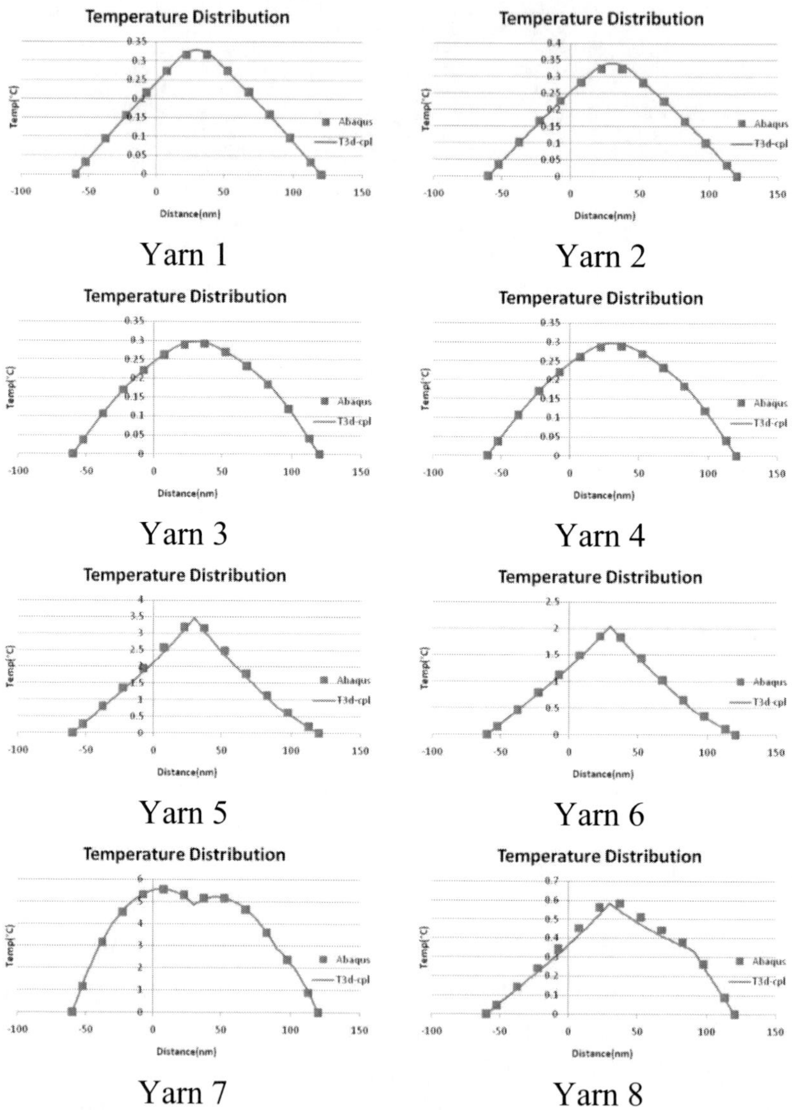

Figure 11. The comparison of all 8 yarns between ABAQUS and Therminator3D with vias and heated yarn 7 in NCF case.

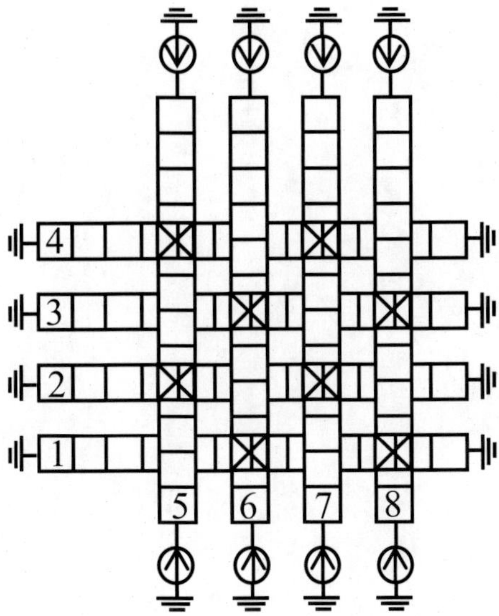

Figure 12. Woven fabric composites represented by a two-layer interconnect grid with vias showing approximate segmentation into connected resistors. The “X” indicate the yarn have vias presented and woven. All endpoint temperatures are set to T = 0°C.

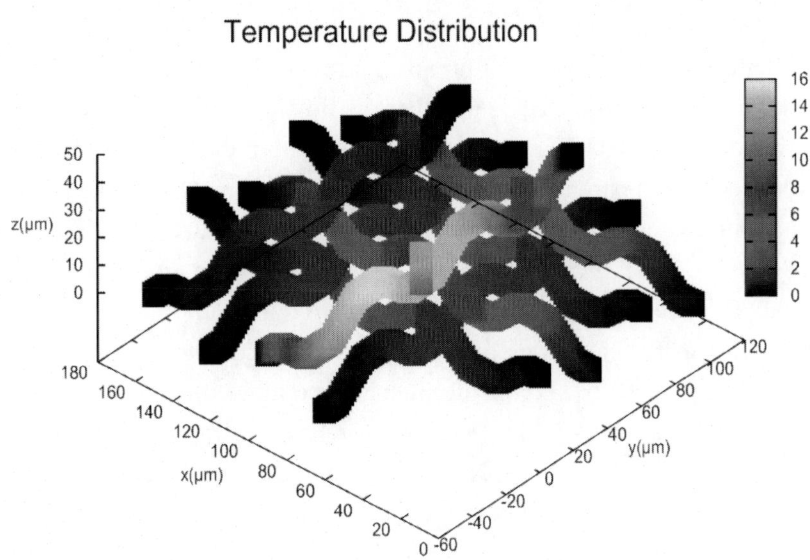

Figure 13. Yarn temperature distribution for the network shown in Figure 12, computed by Therminator3D.

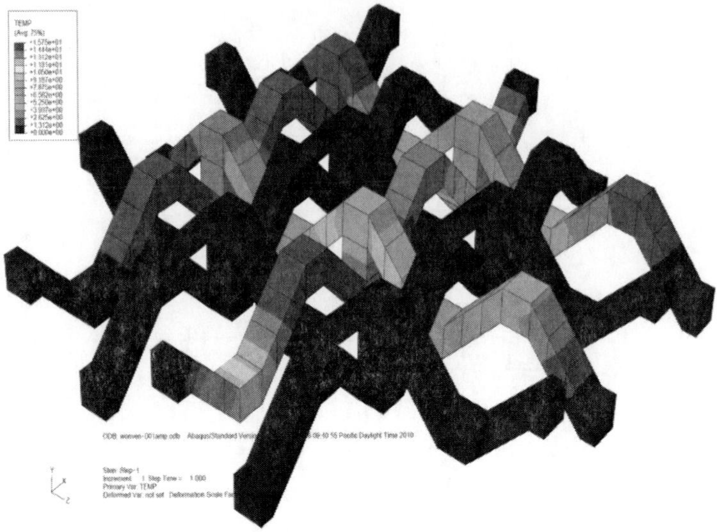

Figure 14. Yarn temperature distribution for the network shown in Figure 12, computed by Therminator3D.

To further evaluate differences between two simulations by ABAQUS and Therminator3D, we listed the temperature profiles of all 8 yarns in Figure 15. As a second test case, we applied the same current to yarn 7. Those temperature distributions on all 8 yarns have been shown in Figure 16. In the woven fabric case, since there is no simple method to compute the exact coupling data, we used a calibration with the results of ABAQUS to estimate the G^{lat} parameters (Figure 5). The Therminator3D results of all yarns are in a good agreement with the simulation of ABAQUS, except for yarn 7 in the first example (Figure 15) and for yarn 5 in the second example (Figure 16). The reason those yarns do not match well the temperature distribution lines of ABAQUS is the complexity of coupling behaviors estimated in Therminator3D. For example, the Therminator3D temperature of yarn 7 is colder than that of ABAQUS in the case of heated yarn 5 (Figure 15). In fact, in Therminator3D, there is no coupling data for yarns 5 and 7 to see each other. In the future work, it is worth continuing to derive more comprehensive coupling formulas in order to obtain more accurate results from Therminator3D. In Table 4, we list the data of CPU time and memory required by Therminator3D and ABAQUS for the performed woven fabric simulations. The results clearly show that Therminator3D has higher speed and memory saving capabilities compared to the ABAQUS simulation. The CPU time for the simulation of Therminator3D is 0.06 sec. The CPU time for the simulation of ABAQUS is 8.1 sec.

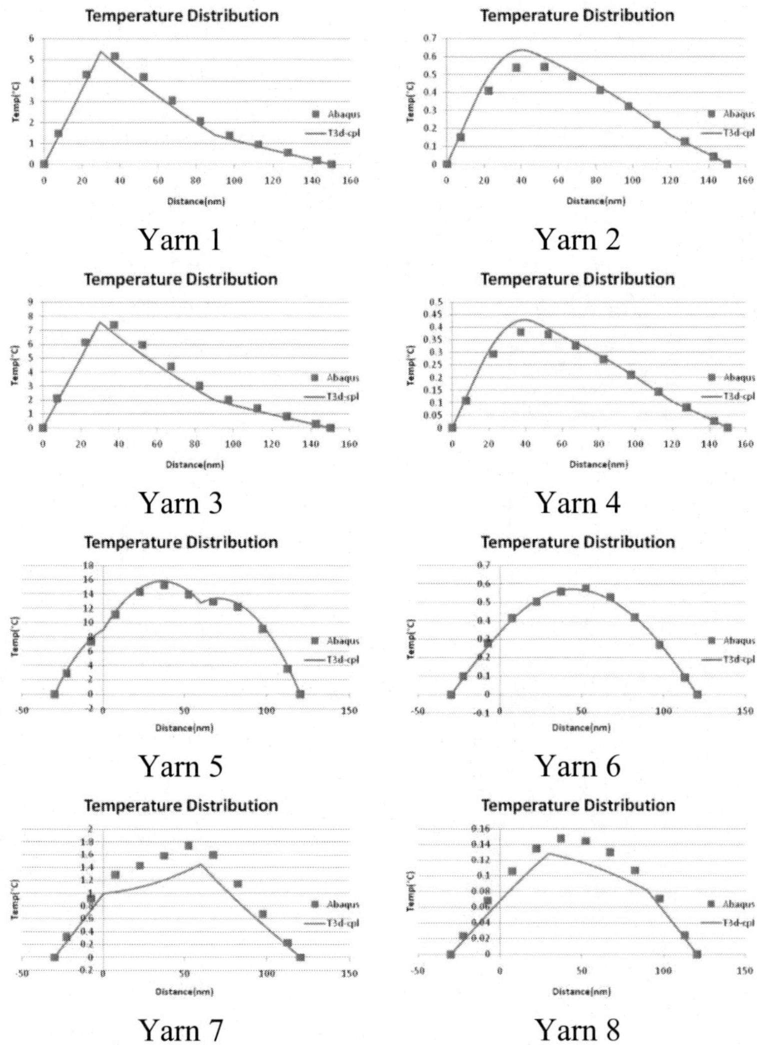

Figure 15. The comparison of all 8 yarns between ABAQUS and Therminator3D with vias and heated yarn 5 in woven case.

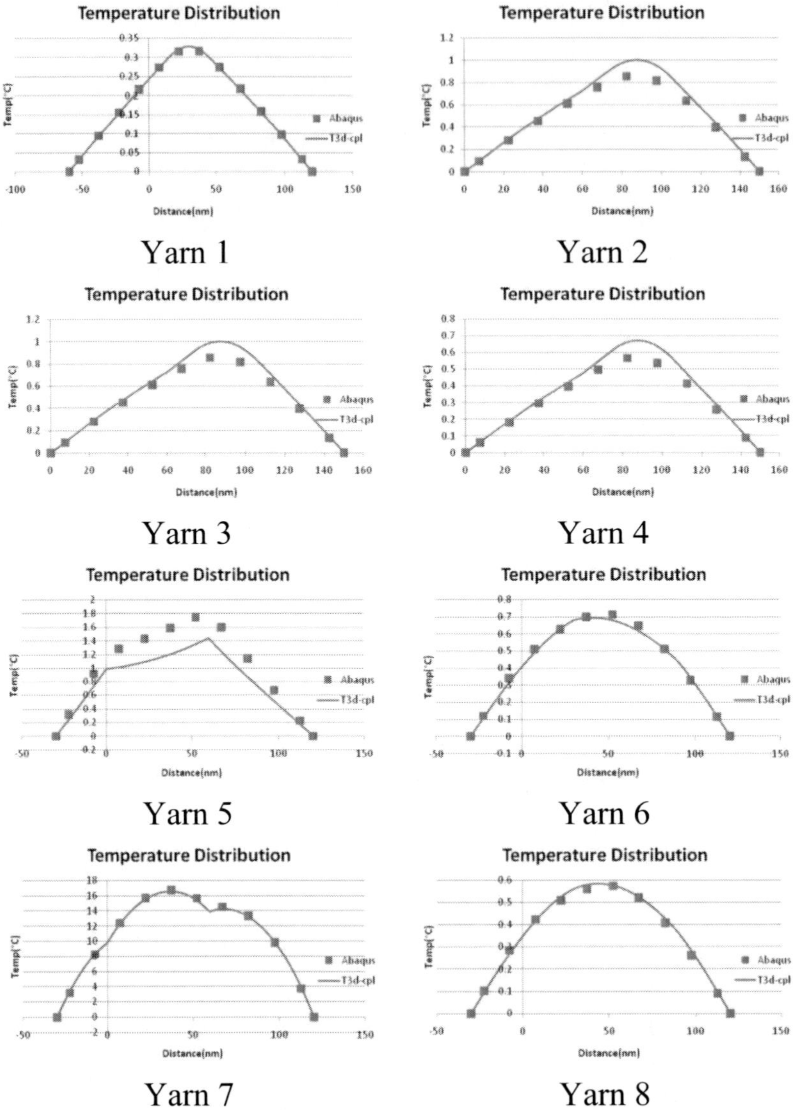

Figure 16. The comparison of all 8 yarns between ABAQUS and Therminator3D with vias and heated yarn 7 in woven case.

The simulation time of woven fabric network is faster than that of non-crimp fabric network on both T3D and ABAQUS tools. The main reason is that the coupling effect of woven fabric network increases the convergence of thermal conduction due to the yarn close to boundary condition such as the ground. In woven fabric network, the values of G_1^{lat} and G_2^{lat} are almost the same in the symmetric structure. In the non-crimp fabric network, the value of G_1^{lat}, which is close to ground, is bigger than G_2^{lat} which is far from the grund. The value of $G_{1,2}^{lat}$ in the woven fabric network is larger than its value in the Non-Crimp Fabric network. The calculation value of equation **(28)** from Woven Fabric Network is closer to 1 compared to its value from the Non-Crimp Fabric Network. It means that Woven Fabric Nework converges faster than Non-Crimp Fabric Network for fixed-point algorithm. The derivations of convergence of Therminator3D, Liao et al. 2010 [19], conclude that the Therminator3D program presents the nature of property without hypothesis assumptions.

3.3. Large-Scale Fabric Network

It is expected that Therminator3D has high capability to simulate large-scale networks. In this section, we use two examples to verify this. The first example shown in Figure 17 is a 20 (columns) by 20 (rows) network with vias. The dimension of each yarn is 10x10x400 μm (WxDxL). The material properties used in this study are given in Table 2. The boundary condition is 0°C temperature on substrate and a current of 0.001 AMP is applied to yarn 3. This network has 1680 nodes including vias for Therminator3D and 45538 nodes for ABAQUS. The CPU time of simulation of Therminator3D was found to be 181.19 sec (including 179 seconds to form the symbolic solution and only 2 seconds for performing the numerical solution). The CPU time of simulation of ABAQUS was 391.70 sec. The symbolic solution in Therminator3D took most of the CPU time before entering iteration loops of numerical solution. The number of iterations in Therminate3D was 35 iterations (in ABAQUS, it was 2 iterations). However, the speed of iterations of Therminator3D was very fast, each iteration only took 0.0571 sec, compared to ABAQUS which took 185.85 sec.

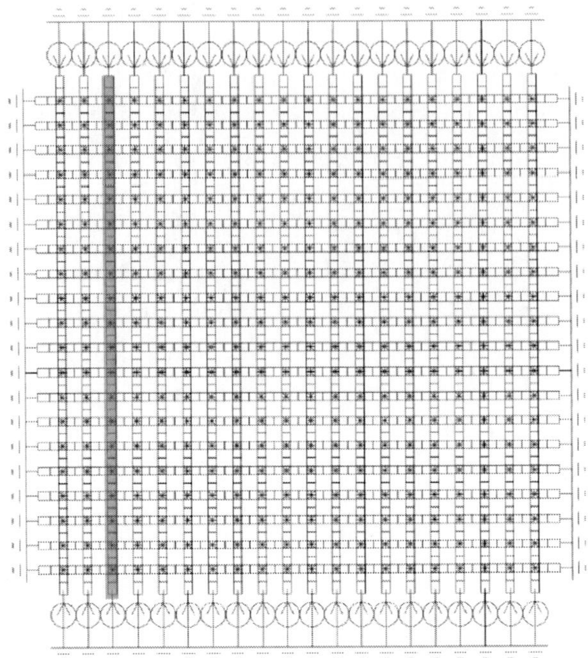

Figure 17. The layout of a 20 (columns) by 20 (rows) large scale network with vias (X) with a heated yarn (red line). The temperature of substrate is 0°C. The porosity of *SiO_2* is considered to be 0%.

The second example shown in Figure 18 is a 20 (columns) by 20 (rows) network without vias and a temperature of 0°C is applied to the ends of all yarns and the substrate. It has 1680 nodes without vias for Therminator3D and 45538 nodes for ABAQUS. The CPU time of Therminator3D is 181.18 sec, but the CPU time of ABAQUS is 325.40 seconds. When compared to the previous example, the CPU time of Therminator3D is not much affected by the presence or absence of vias. It also shows that Therminator3D's iterations only require O(N) additional operations and do not correlate with CPU time.

Figure 19 shows the simulation results of the Therminator3D program and Figure 20 shows the simulation results of the ABAQUS program. From those two figures, the temperature distributions of T3D and ABAQUS are consistent. Table 4 shows that Therminator3D could reach the purpose of fast simulation on large scale networks. In these large-scale simulations, we found that the order of memory of the worst case (which has vias connected with columns and rows) to form the L and U matrices is near to N^3. The reason that such a large memory is required while the size of networks increased, is the solver of the Crout algorithm to form the linked-list hash-tables of L and U matrices. However, the iteration time of the numerical solution is hardly affected by the scale of network.

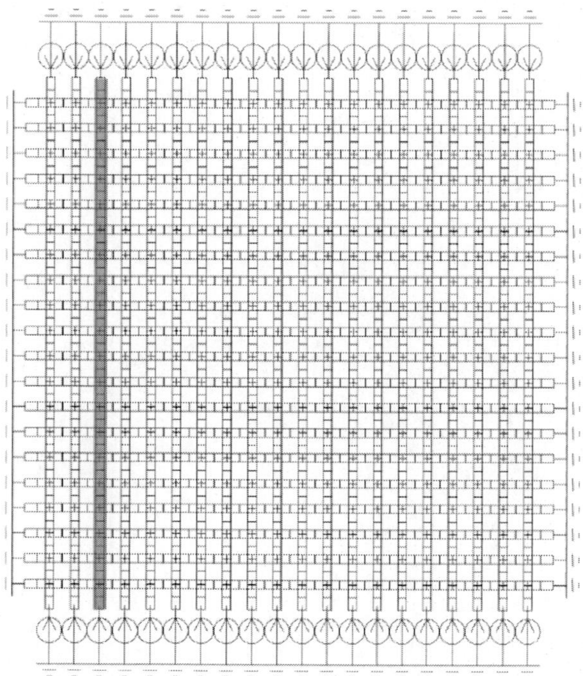

Figure 18. The layout of 20 (columns) by 20 (rows) large scale network without vias and with a heated yarn. The boundary condition of temperature of 0°C is applied to both ends of all yarns and the substrate. The porosity of SiO_2 is considered to be 0%.

The symbolic solution procedure and the incorporation of analytical heat solution over each interconnect segment allows the net-based approach in the Therminator3D simulation to efficiently address the equivalent resistance circuits that incorporate the thermal transport of interconnects. From examples performed in this chapter, it can be concluded that the speed and accuracy of Therminator3D are very satisfactory. The examples also showed that Therminator3D can solve large scale simulations in a short time, and the consumed CPU time is not significantly affected by the iterations required for nonlinear problems. It is also able to deal with several boundary conditions problems. The analytical model and network structure mean Therminator3D uses fewer nodes, and thus less memory than FEA, while offering a competitive accuracy. In the FEA method, the dielectric material is actually included in the CAD model and meshed, whereas in Therminator3D no additional elements are used for the dielectric (matrix).

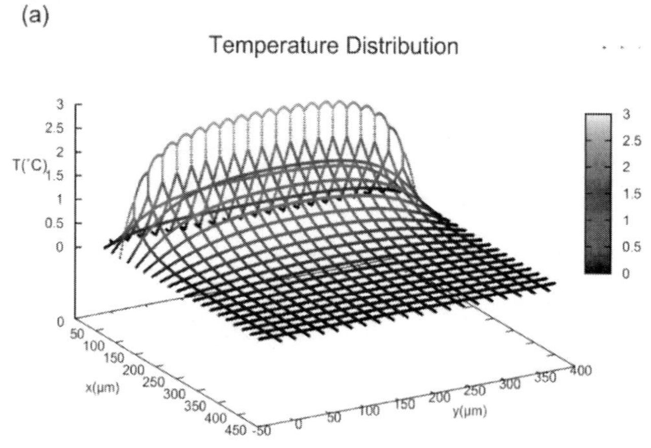

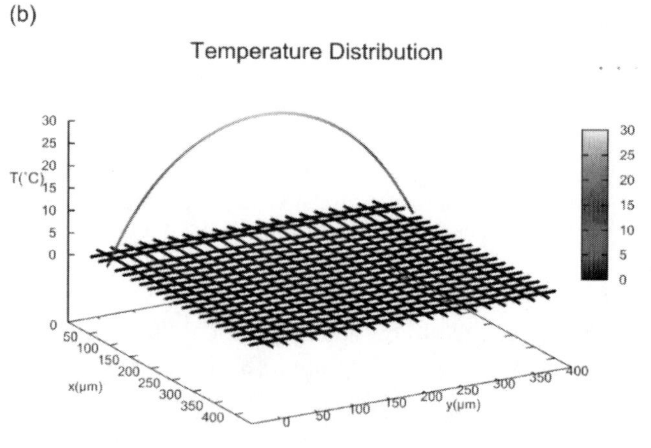

Figure 19. The simulation results of Therminator3D for a 20x20 net. (a) Layout as Figure 17. (b) Layout as Figure 18.

(a)

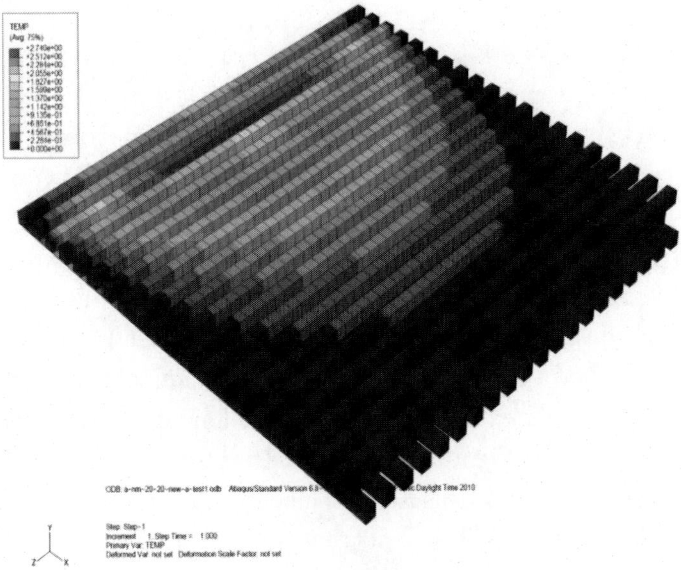

(b)

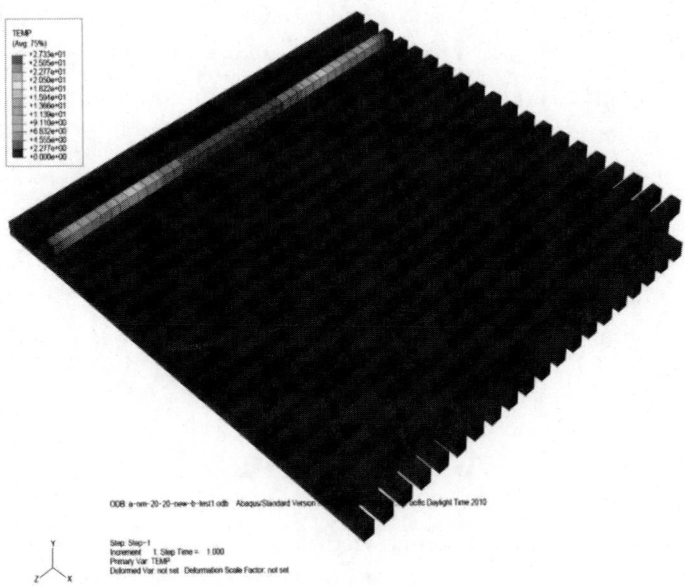

Figure 20. The simulation results of ABAQUS for a 20x20 net. (a) Layout as Figure 17. (b) Layout as Figure 18.

Conclusion and Future Work

A new network-based temperature simulation program, Therminator3D, was successfully developed. This program employs a modified RC network which is developed from electromigration reliability verification concept. Performance of the proposed Therminator3D network-based simulation appeared to be a fast and accurate thermo-electrical simulation tool through cases studies of this paper. The main characteristics of Therminator3D can be summarized as follows: (i) Network-based approach in Therminator3D efficiently addresses the equivalent resistance circuits that incorporate the thermal transport of fiber. (ii) It can solve large textile composite simulations in a short time and the consumed time is not significantly affected by the iterations required for nonlinear problems. (iii) It uses fewer nodes and thus less memory than FEA by analytical models and network structures. In FEA, the dielectric/matrix material is actually included in the CAD model and meshed, whereas in Therminator3D no additional elements are used for the dielectric/matrix. (iv) It has a good potential to develop several other mechanical simulations on material behaviors, such as strain-stress response which is normally obtained by FEA, and multi-physic problems which often require higher computational resources.

To achieve these future goals, there are two phases in future: (i) Develop more advanced coupling formulas to include layout variables such as the angle of yarn, distance, effective properties, etc. (ii) Modify the Therminator3D code with algorithm, which needs less memory consumption in symbolic solution, to increase its speed even further.

Acknowledgment

The authors wish to acknowledge the financial support from the Natural Sciences and Engineering Research Council (NSERC) of Canada.

References

[1] Labun A, Jagjitkumar K. Rapid detailed temperature estimation for highly coupled IC interconnect. *IEEE Trans Comput Des Integr Circuits Syst*. 2008;27(10):1840-1851.

[2] Jain LK, Dransfield K a., Mai YW. On the effects of stitching in CFRPs—II. Mode II delamination toughness. *Compos Sci Technol*. 1998;58(6):829-837.

[3] Hind S, Raizenne D, Robitaille F. No Title. In: *Parametric Unit Cell Modelling of the Effective Transverse Thermal Conductivity of Carbon Plain Weave Composites*. Vol; 2006.

[4] Nordlund M, Lundström TS, Frishfelds V, Jakovics A. Permeability network model for non-crimp fabrics. *Compos Part A Appl Sci Manuf*. 2006;37(6 SPEC. ISS.):826-835.

[5] Nordlund M, Frishfelds V, Jakovics A. Application of Permeability Network Model To Non-Crimp Fabrics. *Network*. 2004;c:1-5.

[6] Ning QG, Chou TW. A Closed-Form Solution of the Transverse Effective Thermal Conductivity of Woven Fabric Composites. *J Compos Mater*. 1995;29(17):2280-2294.

[7] Xu J, Wirtz R a. In-Plane Effective Thermal Conductivity of Symmetric, Diamond-Weave Screen Laminates. *J Electron Packag*. 2005;127(3):353.

[8] Peng X, Lu X, Balendra R. FE simulation of the blanking of electrically heated engineering materials. *J Mater Process Technol*. 2004;145(2):224-232.

[9] Zhao Q, Hoa S V. Thermal Deformation Behavior of Triaxial Woven Fabric (TWF) Composites with Open Holes. *J Compos Mater*. 2003;37(18):1629-1649.

[10] Xiong JJ, Shenoi RA, Gao J. An analytical model to predict residual thermal stress in 2D orthogonal plain weave fabric composites. *Int J Solids Struct*. 2009;46(9):1872-1883.

[11] Li L, Kim SM, Song SH, et al. Finite element modeling and simulation for bending analysis of multi-layer printed circuit boards

using woven fiber composite. *J Mater Process Technol*. 2008;201(1):746-750.

[12] Clement JJ, Riege SP, Cvijetic R, Thompson C V. Methodology for electromigration critical threshold design rule evaluation. *IEEE Trans Comput Des Integr Circuits Syst*. 1999;18(5):576-581.

[13] Teng C, Cheng Y, Rosenbaum E, Kang S. iTEM : *A Chip-Level Electromigration Reliability Diagnosis Tool Using Electrothermal Timing Simulation*. 1996:172-179.

[14] Chiang TY, Saraswat KC. Closed-form analytical thermal model for accurate temperature estimation of multilevel ULSI interconnects. *2003 Symp VLSI Circuits Dig Tech Pap (IEEE Cat No03CH37408)*. 2003:275-278.

[15] Tsui B, Yang C. Anisotropic thermal conductivity of nanoporous silica film. *Electron Devices, IEEE*. 2004;51(1):20-27.

[16] Tiwari D, Basu B, Biswas K. Simulation of thermal and electric field evolution during spark plasma sintering. *Ceram Int*. 2009;35(2):699-708.

[17] Ishikawa T, Chou TW. Stiffness and strength behaviour of woven fabric composites. *J Mater Sci*. 1982;17(11):3211-3220.

[18] Donchin BDR, Fischer TC, Fox TF, Peng V, Preston P, Wheeler WR. The NVAX CPU Chip: Design Challenges, Methods, and CAD Tools By Dale R. Donchin, Timothy C. Fischer, Thomas F. Fox, Victor Peng, Ronald P. Preston, and William R. Wheeler 1. *IEEE*. 1992:24-37.

[19] Liao CS. *A Network Approach for Thermo-Electrical Modelling: from IC Interconnects to Textile Composites*. 2010.

[20] Hibbitt K and SI. *ABAQUS Theory Manual*.; 2004.

[21] Labun A. Thermally-coupled IC interconnect networks. *2008 IEEE 14th Int Mix Sensors, Syst Test Work*. 2008:1-5.

In: Polymer-Matrix Composites ISBN: 978-1-53613-398-1
Editors: Elias Rice and Beth Sparks

Chapter 7

AN INVESTIGATION OF THE EFFECTS OF RAMP TEMPERATURE DEGRADATION OF COPPER (II) OXIDE ADDED POLYETHYLENE/ETHYLENE-VINYL ACETATE COMPOSITE

Tze-Ming Chee[1], Lee Tin Sin[1,*], Soo-Tueen Bee[1,†], Tiam-Ting Tee[1] and A. R. Rahmat[2]

[1]Department of Chemical Engineering, Lee Kong Chian Faculty of Engineering and Science, Universiti Tunku Abdul Rahman, Bandar Sungai Long, Jalan Sungai Long, Kajang, Selangor, Malaysia

[2]Department of Polymer Engineering, Faculty of Chemical Engineering, Universiti Teknologi Malaysia, UTM Skudai, Johor, Malaysia

* Corresponding Author: Tel: +60 3 9086 0288; Fax: +60 3 9019 1989. Email: direct.tinsin@gmail.com.

† Corresponding Author E-mail: direct.beest@gmail.com.

ABSTRACT

This study was conducted to investigate the effects of copper (II) oxide (CuO) composition on the degradability of high density polyethylene/ethylene vinyl acetate (HDPE/EVA) blends. The content of CuO incorporated was varied from 0 – 5 phr and the compounded composites were subjected to ramp-120°C and ramp-100°C degradation. The tensile results showed that the all composites showed deterioration of mechanical properties after the degradation process. In terms of thermal properties, the differential scanning calorimatry (DSC) analysis suggested that both the melting temperature (T_m) and melting heat (ΔH_m) increased after 6 days degradation and the values were observed to reduce again when the degradation prolonged. Similar trend was observed to the temperature at 5% weight loss ($T_{5\%}$) of composites for the thermogravimetric analysis (TGA). The physical properties of the composites have been investigated through the X-ray diffraction (XRD) analysis. It was observed that both degree of crystallinity and crystallites sizes of the composited increased after the ageing operation. Finally, the impacts of the CuO composition on the degradability of the composites have been verified through the SEM analysis.

Keywords: copolymers, degradation, polyolefins, properties and characterizations

1. INTRODUCTION

The incorporation of inorganic fillers or reinforcement fibers into polymer blends is a common practice in the polymer industry of today [1]. The addition of these fillers is basically aims to alter the properties of the polymer blends. For instance, certain inorganic fillers acted as reinforcing agents and enhanced the tensile strength of the blends while certain inorganic fillers acted as degradants which accelerated the degradability of the blends. One of the most common types of inorganic fillers is transition metal oxide such as zinc oxide or titanium oxide. For instance, Pavličević et al. [2] reported that the addition of zinc oxide significantly reduced the extent of phase separation of polycarbonate/polyurethane blends where the maximum tensile strength took place at 1 wt% of filler content. On the other hand, Nakane et al. [3] reported that the incorporation of titanium dioxide has potentially improved the thermal properties of poly(vinyl butyral) blend.

Cupric oxide/copper (II) oxide (CuO) is an inorganic compound which has been used in various applications including gas sensors, catalysis, batteries, field emission emitters and high-temperature superconductor [4]. The good physical properties of CuO such as electron correlation effects and spin dynamics have potentially gained the attention of researchers in developing its diverge applications [5]. As CuO is one of the commercially affordable and readily available inorganic filler, it has also been widely incorporated in various polymer blends to enhance their properties. For instance, Hong and Wang [6] established that the incorporation of copper (II) oxide (with 0.1 wt%) significantly promoted the thermal-oxidative degradation of epoxy-based resins.

Polyethylene/ethylene-vinyl acetate (PE/EVA) blends are well-established polymer blends in the polymer industry. The applications of the blends include the manufacture of multilayer packaging, shrinkable films as well as wire and cable coating [7 – 10]. Meanwhile, the presence of high vinyl acetate (VA) content in PE/EVA blends also improves the properties of the blends to suit the diverge applications of commercial products including the synthesis of tubes, tires, adhesives and so on [11,12]. This potentially marked down the high demand of PE/EVA blends from the polymer industry of today. In this study, small quantity of CuO was incorporated into PE/EVA blends to determine the degradability of the blends so that without compromising the mechanical properties of the blends. The target of this study is to determine the action of traces of CuO which is common present as impurities either from the sources of additives or corroded machineries body. In the past, the degradability of the various polymer blends have been studied by the researchers through the thermal degradation techniques by applying elevated temperature to the polymer blends. Many researchers have reported that the metal/polymer blends showed obvious degradation when they have undergone isothermal degradation. However, it is crucial to note that many applications of

PE/EVA involve ramp heating process such as containers, plastic body in electrical appliances with heating function as well as wire/cable. Although a ramp heating process is assumed to have milder effects to the polymer blends compared to the isothermal process, the application of heat energy would still lead to some extent of degradation to the polymer products. Up to date, the study of cyclic thermal effects are left unattended and this information can be important especially for plastic products manufacturers to meet the high demand of quality of consumers for longer last subsequently reduce the expenses for maintenance and replacement.

2. EXPERIMENTAL

2.1. Materials and Sample Preparation

In this study, three main materials namely, copper (II) oxide (CuO) (Sigma-Aldrich Malaysia, analytical reagent grade), high density polyethylene (HDPE) (Polyethylene Sdn. Bhd. Malaysia, grade ETILINAS HD5301AA), and ethylene-vinyl acetate (EVA) (USI Corporation, Taiwan, grade UE629 containing 18 wt.% vinyl acetate), were used. These three materials were formulated where both HDPE and EVA contents were fixed at 100 phr and the content of CuO was varied from 0 to 5 phr. All formulated materials were compounded by using Nanjing Giant SHJ-20 Twin-Screw Extruder at average screw rotational speed of 5 rpm and process temperature of 180°C. After the extrusion process, the composites were then compressed into 1 mm × 15 cm × 15 cm sheets by using Lotus Scientific L5-11009 hot presser. Thereafter, dumbbell-shaped sample specimens that were based on ASTM D1822 standard were obtained from the compressed sheets by using Cometech QC-603A pneumatic tensile cutter. The sample specimens were then subjected to ramp temperature degradations (from 30 to 100°C and from 30 to 120°C) in Memmert Universal Oven for 3, 6, 9 and 12 days. The temperature profiles of the degradation method are shown in Figure 1.

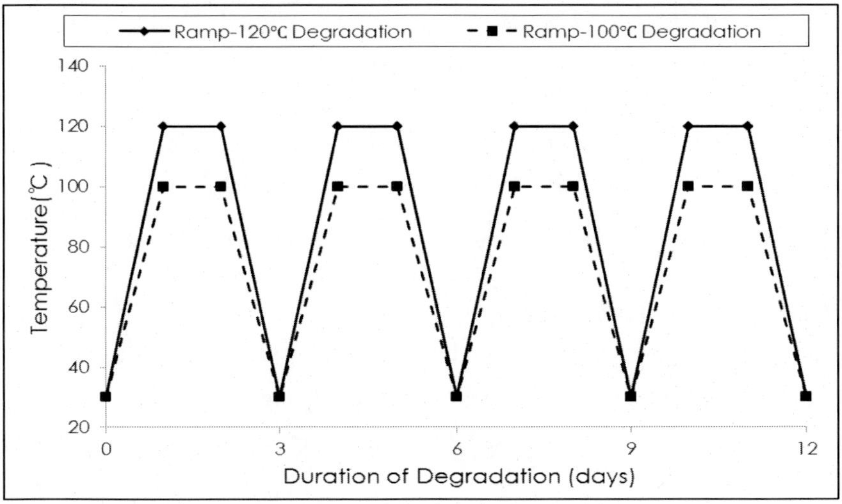

Figure 1. Temperature Profiles for the Ramp Temperature Degradation.

2.2. Characterization Techniques

2.2.1. Tensile Test

The mechanical properties of the sample specimens were evaluated by the tensile test, using Instron Universal Testing Machine 5582 series IX tensile tester. The measurement was carried out at room condition with constant load cell of 2 kN and crosshead speed of 50 mm/min. Two tensile parameters, namely tensile strength and elongation at break, for each sample specimen were obtained directly from the tensile datasheet that was generated by Bluehill software. For each formulated sample, five determinations of each parameter were carried out and the result was taken based on the average values of the determination.

2.2.2. Differential Scanning Calorimetry

In order to characterize the melting behaviors of the composites, the un-degraded composites and the composites that have been subjected to 6 and 12 days degradation were subjected to differential scanning calorimetry (DSC) analysis, using Mettler-Toledo DSC823E/200 analyzer.

In each measurement, 5 – 8 mg of sample was heated dynamically from 30 to 200°C at constant heating rate of 10°C/min. Dry nitrogen gas at constant volumetric flowrate of 20 ml/min was applied for purging purpose throughout the entire measurement. Both melting temperature (T_m) and melting heat (ΔH_m) of each analyzed sample were obtained direct from the DSC thermogram that was generated by Star[e] software.

2.2.3. Thermogravimetric Analysis

In order to characterize the thermal stability of the composites, the un-degraded composites and composites that have been subjected to 6 and 12 days degradation were subjected to thermogravimetric analysis (TGA), using Mettler-Toledo TGA/SDTA851[e] analyzer. In each measurement, 5 – 8 mg of sample was heated from 25 to 800°C at constant heating rate of 20 K/min. Dry nitrogen gas at constant volumetric flowrate of 20 ml/min was applied for purging purpose. The temperature at 5% weight loss ($T_{5\%}$) for each tested samples were obtained from the TGA thermogram that was generated by the Star[e] software.

2.2.4. X-Ray Diffraction

In order to characterize the physical properties of the composites, un-degraded composites and composites that have been subjected to 6 and 12 days degradation were subjected to X-ray Diffraction (XRD) analysis, using Shimadzu XRD 6000 Diffractometer with Cu-Kα radiation. The scattering angle (2θ) of each measurement was varied from 3 to 40° and the scanning rate applied was 1°/min. For the X-ray tube conditions, the voltage and current applied were 40.0 kV and 30.0 mA, respectively. The degree of crystallinity for each sample was obtained directly from the XRD datasheet that was generated by XRD-6000 software. Apart from that, the crystallites sizes for each sample have also been calculated by Scherrer's Equation as shown in Equation 1:

$$L = \frac{k\lambda}{\beta \times \cos\theta} \tag{1}$$

where

L = crystallite size, Å

k = Scherrer Constant

λ = wavelength of Cu-Kα radiation applied (1.542 Å)

θ = angle between X-ray beam and the normal axis of the reflecting plane, degree

β = FWHM of the diffraction peak, radian

2.2.5. Scanning Electron Microscopy

In order to characterize the surface morphologies of the composites, un-degraded composites and composites that have been subjected to 6 and 12 days ageing were subjected to scanning electron microscopy (SEM) analysis, using JOEL JSM-3301F SEM unit that was operated at voltage value of 15.0 kV. After the completion of tensile test, the fractured region of the composite was then cut and placed onto a copper stub such that the fractured surface was facing directly to the electron gun. A thin layer of gold was coated onto the entire sample surface and the coated sample was then subjected to the analysis. Microphotograph with magnification of 1,000 times was taken for each analyzed sample.

3. Results and Discussion

3.1. Tensile Test

For the tensile properties of the composites, every sample showed typical stress-strain behavior of a rubbery material throughout the test. For the un-degraded composites, it was observed that composites with 1 and 2 phr of CuO showed lower tensile strength as compared to that of unfilled HDPE/EVA blend. When the incorporation of CuO was increased to 3 phr, the composite showed maximum a tensile strength, as shown in Figure 2(a) and 2(b). This was attributed to the good quality of CuO particle dispersion which significantly reinforced the polymer composite [13]. When the CuO content was further increased to 4 phr and above, the tensile strength of the composites was observed to reduce. This was due to the agglomeration of CuO particles where the CuO-CuO particles interaction was stronger than that of CuO particles-polymer matrix interaction [14]. For composites that have been subjected to ramp-120°C degradation, it was observed that all composites showed drastic reduction in tensile strength after 3 days degradation as shown in Figure 2(a). As the duration of ageing was prolonged to 3 days and above, no remarkable changes were observed. Apart from that, the variation of tensile strength after thermal ageing was observed to be independent to the composition of CuO.

On the other and, composites that have been subjected to ramp-100°C degradation were also observed to show single step reduction in tensile strength as shown in Figure 2(b). The reduction in tensile strength was attributed to the chain scissioning process where certain parts of the polymer chains have been decomposed during the course of thermal ageing. As the duration of thermal ageing was prolonged to 6 days and above, the trend of the tensile strength was observed to increase with CuO content. This implies that the polymer chains have undergone recrystallization and the reinforcement effects became more significant at higher CuO content.

Elongation at break denotes the maximum elongation of the composite before failure during the tensile extension [13]. By referring to Figure 3(a) and 3(b), it can be observed that composites with 1 and 2 phr of CuO showed lower elongation at break as compared to that of unfilled HDPE/EVA blend before the degradation. This could be due to the strong adhesion effects between polymer matrix and CuO particles which restricted the chain movement of the blends. Maximum elongation at break was observed when the CuO content was at 3 phr. This could be attributed to the good particle distribution that causing the small distance between the aggregates and enhancing the plastic deformation [15]. When the CuO content was further increased to 4 phr and above, the composites possessed minor reduction in elongation at break. This could be due to the reduction in interfacial adhesion effects between polymer matrix and CuO particles as the CuO particles have formed larger clusters structures.

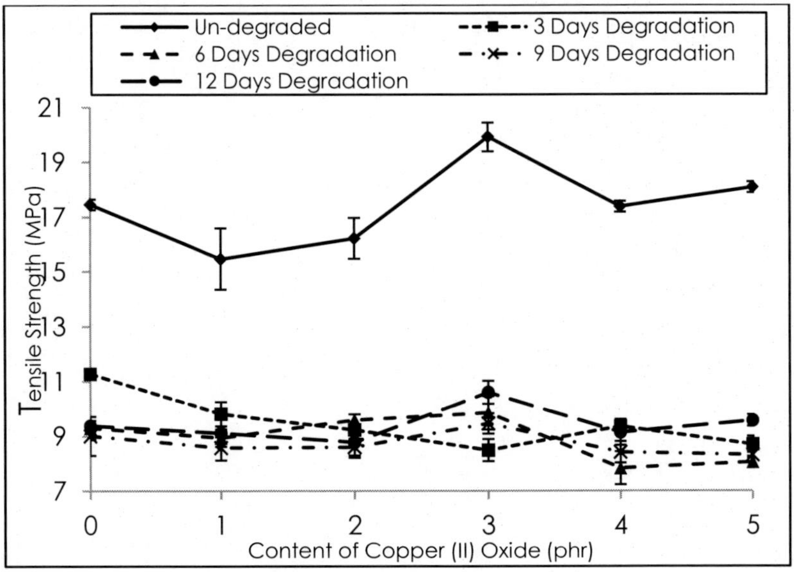

(a)

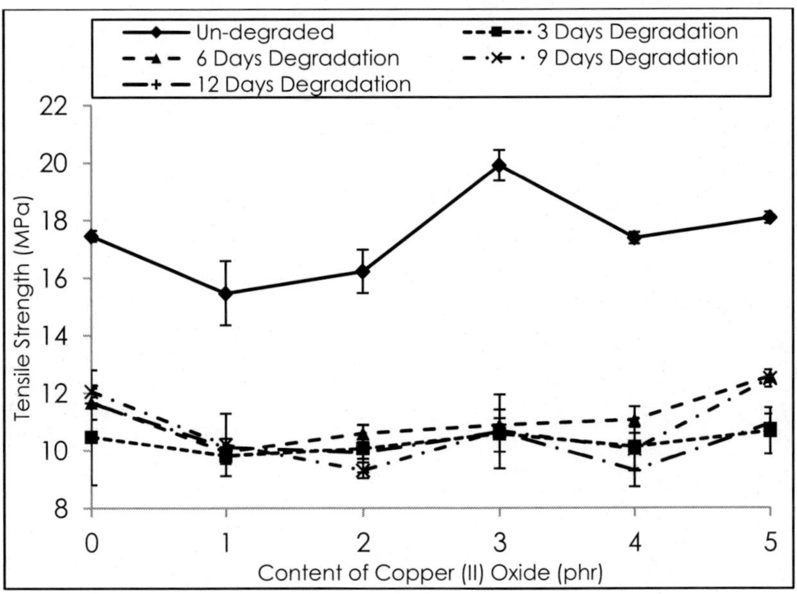

(b)

Figure 2. Tensile Strength for HDPE/EVA Blends at Various CuO Compositions for (a) Ramp-120°C Degradation and (b) Ramp-100°C Degradation.

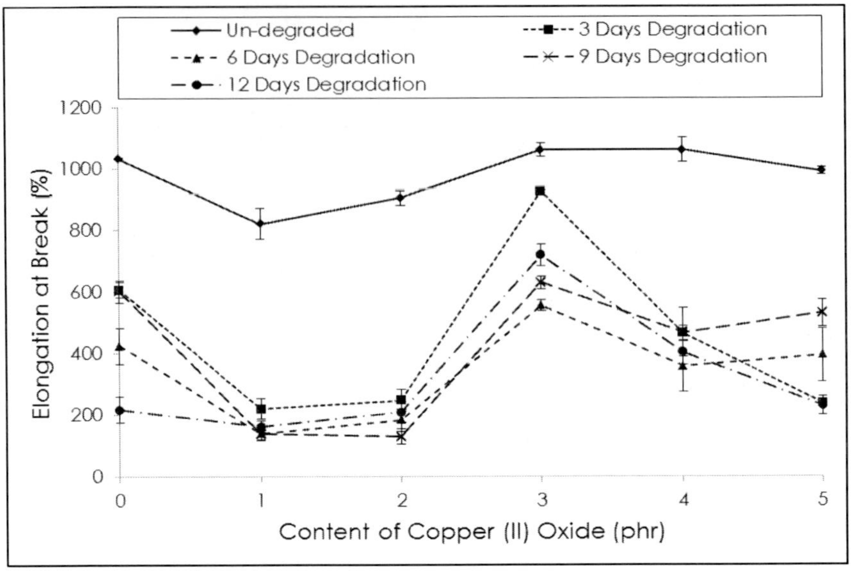

(a)

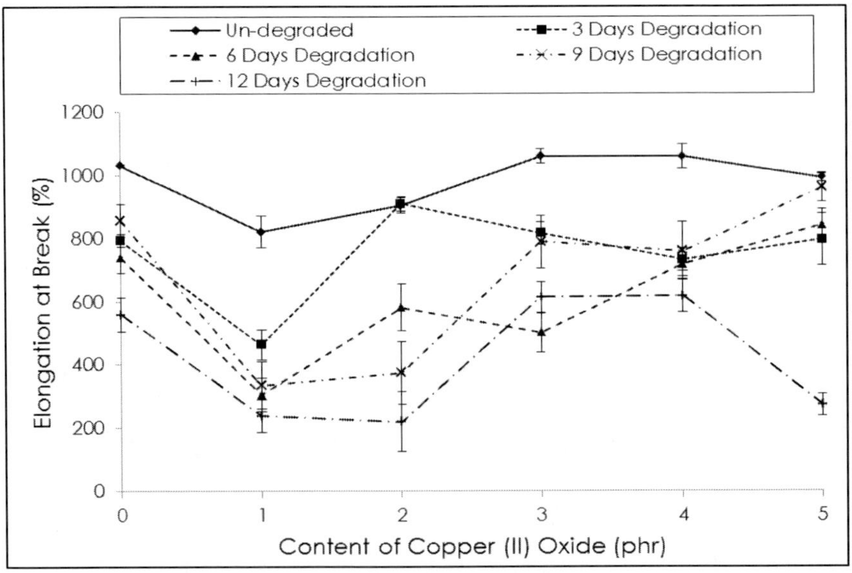

(b)

Figure 3. Elongation at Break for HDPE/EVA Blends at Various CuO Compositions for (a) Ramp-120°C Degradation (b) Ramp-100°C Degradation.

When the composites were subjected to ramp-120°C degradation, the values for elongation at break for all composites were observed to decrease. This was attributed to the chain scissioning process where the polymer chains have been discontinued with each other during the course of thermal degradation. Apart from that, it can also be observed from Figure 3(a) that trends of elongation at break after the degradation process are almost similar to that of un-degraded condition where the value of elongation at break for composite with 3 phr of CuO remains at maximum. This implies that the mechanisms of plastic deformation remain almost similar to that of un-degraded condition after the prolonged thermal degradation. Furthermore, the reduction in elongation at break for composite with high CuO content (4 and 5 phr) was observed to be more profound after the ramp-120°C degradation. This could be due to the participation of CuO particles in inducing the chain scissioning process by releasing active radicals at elevated temperature [16].

For composites that have been subjected to ramp-100°C degradation, it can be observed from Figure 3(b) that only composite with 1 phr of CuO showed the similar trend to the un-degraded condition and ramp-120°C degradation. As the CuO content increased progressively, the trends of the elongation at break for the composite were observed to increase. In addition, the values of elongation at break for composites with 2 to 5 phr of CuO were observed to vary inconsistently as the duration of thermal ageing prolonged. This could be explained through the free radical reaction where polymer free radicals that were produced at elevated temperature (the periods where temperature was increased from 30 to 100°C and kept at constant 100°C) have recombined simultaneously with each other when the temperature was decreased from 100 to 30°C. This will certainly result in the formation of polymer chains that are with high bonding spectrum.

3.2. Differential Scanning Calorimetry

In this study, the melting behaviors of the composites were characterized through the investigation of the T_m and ΔH_m of composites.

Both the parameters were obtained directly from the melting stage of the DSC thermograms, where T_m was represented by the peak temperature and ΔH_m was represented by the area under the curve at melting stage. By referring to Tables 1 and 2, both T_m and ΔH_m for the un-degraded composites were observed to reduce with the increment of CuO content. This was attributed to the decrease in degree of crystallinity where less orderly arranged crystal structures were needed to be overcame at high CuO content.

Table 1. Melting Temperature for Un-degraded Composites and Composites that have been subjected to Ramp-120°C and Ramp-100°C Degradation at Various CuO Compositions

Content of CuO (phr)	Melting Temperature, T_m (°C)				
	Un-degraded	Ramp-120°C Degradation		Ramp-100°C Degradation	
		6 Days Degradation	12 Days Degradation	6 Days Degradation	12 Days Degradation
1	130.92	134.11	135.52	130.17	131.66
3	130.20	134.53	134.07	130.73	131.25
5	129.69	133.71	133.69	130.93	133.43

Table 2. Melting Heat for Un-degraded Composites and Composites that have been subjected to Ramp-120°C and Ramp-100°C Degradation at Various CuO Compositions

Content of CuO (phr)	Melting Heat, ΔH_m (J/g)				
	Un-degraded	Ramp-120°C Degradation		Ramp-100°C Degradation	
		6 Days Degradation	12 Days Degradation	6 Days Degradation	12 Days Degradation
1	76.10	76.19	72.31	83.91	83.11
3	74.93	73.57	72.99	85.80	82.48
5	68.61	70.06	69.09	81.95	82.04

When the composites were subjected to ramp-120°C degradation for 6 days, the T_m for all the composites were observed to increase, as shown in Table 1. This was reasonable as certain parts of the polymer chains (with vinyl acetate groups) have been decomposed and vaporized during the course of thermal degradation. Apart from that, the polymer chains have also gained sufficient energy to undergo recrystallization during the course of thermal degradation and formed more orderly arranged crystal structures. This will certainly lead to the increase in ΔH_m, as shown in Table 2. As the duration of thermal degradation was further prolonged to 12 days, no remarkable changes in T_m were observed. However, the values of ΔH_m were observed to reduce and this indicates the destruction of crystal structures after the prolonged degradation process. In terms of CuO composition, both the T_m and ΔH_m values of the composites were also observed to reduce with the increment of CuO content. This could be attributed to the role of CuO particles in inducing the chain scissioning process where the polymer chains have been degraded into shorter polymer chains after the degradation process.

For composites that have been subjected to ramp-100°C degradation for 6 days, the composites did not possess significant changes in T_m, as shown in Table 1. However, the values of ΔH_m were observed to increase drastically after the ageing operation. This was consistent with the results obtained in the previous section where the polymer free radicals that were produced at elevated temperature have recombined with each other after the heating operation. Furthermore, the increase in ΔH_m could also be attributed to the recrystallization of polymer matrix during the thermal degradation. When the duration of thermal degradation was prolonged to 12 days, the values of T_m were observed to increase. Apart from that, the increase in T_m was seen to be more profound for composite with 5 phr of CuO. On the other hand, the composites were observed to show minor reduction in ΔH_m and this can be attributed to the decrease in degree of crystallinity after the prolonged thermal degradation.

3.3. Thermogravimetric Analysis

In this study, the thermal stabilities of the composites were investigated by the TGA measurement. Throughout the measurement, all tested samples were observed to show two-stage weight losses where the first stage (300 to 400°C) was due to the decomposition of EVA chains and the second stage degradation (400 to 500°C) was due to the decomposition of HDPE chains [17]. In order to characterize the thermal stabilities of the composites, comparison of $T_{5\%}$ has been made. $T_{5\%}$ denotes the temperature where 5% of the total sample weight has been completely combusted during the TGA measurement. By referring to Figure 4(a) and 4(b), it can be observed the composite with 3 phr showed the maximum value of $T_{5\%}$ before the degradation. This could be attributed excellent heat capacity of the well-dispersed CuO particles which absorbed the heat energy reaching the polymer matrix [18]. When the CuO content was further increased to 5 phr, the value of $T_{5\%}$ was observed to reduce. This was attributed to the good thermal conductivity of CuO particles which effectively dissipated the heat energy to the neighboring polymer matrix and causing the decomposition to occur at relatively lower temperature [19].

When the composites were subjected to ramp-120°C degradation for 6 days, the values of $T_{5\%}$ for all tested composites were observed to increase, as shown in Figure 4(a). This observation was consistent with the results obtained in DSC analysis where certain parts of the polymer matrix have been vaporized during the course of thermal degradation. This will lead to the delay in $T_{5\%}$ value as more heat energy was needed to decompose the relatively stronger polymer chains. As the duration of thermal degradation was prolonged to 12 days, the composites were observed to show minor reduction in $T_{5\%}$ values, indicating the formation of more short chain polymers in the composite during the degradation. Furthermore, it can also be observed that the maximum $T_{5\%}$ value took place at 1 phr of CuO content and the values were observed to reduce with the increment of CuO content. This implies that the presence of high CuO content has significantly induced the chain scissioning of polymer matrix during the course of ramp-120°C degradation.

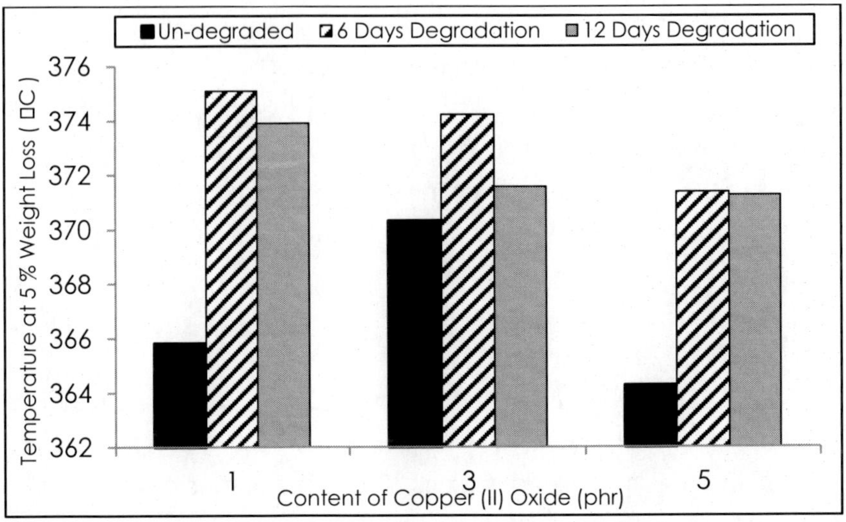

(a)

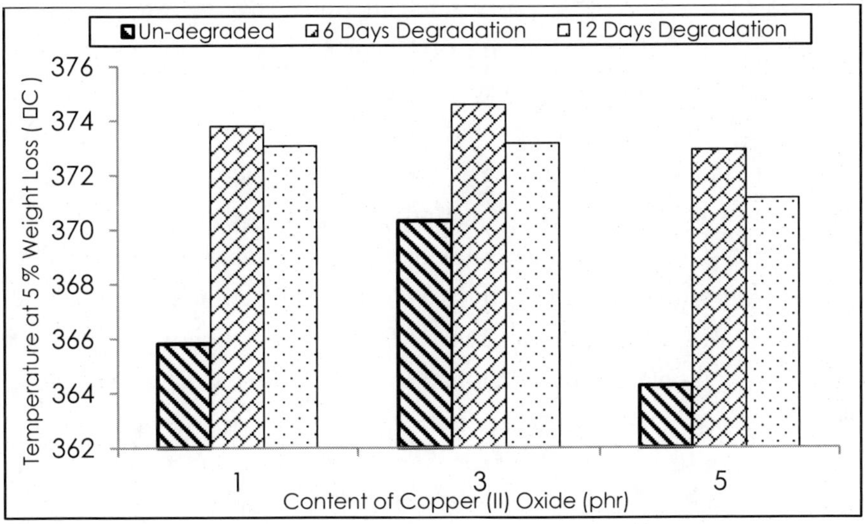

(b)

Figure 4. Temperature at 5% Weight Loss for HDPE/EVA Blends at Various CuO Compositions for (a) Ramp-120°C Degradation and (b) Ramp-100°C Degradation.

For composites that have been subjected to ramp-100°C degradation, the changes in $T_{5\%}$ values were observed to be similar to that of ramp-120°C. However, the maximum value of $T_{5\%}$ was observed to occur at 3 phr of CuO content after 6 and 12 days of thermal degradation. This observation was similar to that of un-degraded and can be explain through the quality of CuO particle dispersion. This also indicates that the CuO composition did not show profound alteration to the decomposition behaviors of the composites after the ramp-100°C degradation.

3.4. X-Ray Diffraction

It was well established that all polymer blends contain numerous amount of tiny crystal structures which are randomly distributed and joined by the intervening amorphous regions [20]. The total amount of crystal structures in the polymer matrix can be represented by the term called degree of crystallinity. Due to the presence of orderly arranged crystal structures that were held by the strong intermolecular forces, composite with high degree of crystallinity will certainly possess good mechanical and thermal properties. In order to characterize the physical properties of the composites, comparisons on the degree of crystallinity have been made. For the un-degraded composites, the degree of crystallinity of the composites were observed to reduce progressively with the increment of CuO content, as shown in both Figure 5(a) and 5(b). This was attributed to the hindering effect that caused by the CuO particles which disturbed the arrangement of the crystal structures [21]. This observation was consistent with the results obtained by Johan and Leo [22] who has investigated the degree of crystallinity of nano-CuO added polyethylene oxide blends.

For composites that have been subjected to both ramp-120°C and ramp-100°C degradation for 6 days, the degree of crystallinity were observed to increase, as shown in Figure 5(a) and 5(b), respectively. This indicates that throughout the 6 days degradation, the polymer chains have gained sufficient energy to undergo recrystallization process and formed orderly arranged structures [23]. As the duration of the degradation was prolonged to 12 days, composites that have been subjected to ramp-120°C degradation showed profound reduction in degree of crystallinity than that of ramp-100°C. This was attributed to the higher degradation temperature which ruptured certain parts of the crystal structures in the composites. In terms of CuO composition, it was observed that the maximum degree of crystallinity for composites that have undergone ramp-120°C degradation took place at 3 phr of CuO content. This could be due to the mobilization of polymer chains which minimized the hindering effects that caused by the well dispersed CuO particles. On the other hand, the trends for the variation of degree of crystallinity for composites that have been subjected to ramp-100°C degradation were seen to be similar to that of un-degraded condition, indicating the existence of hindering effects caused by the CuO particles.

Apart from the degree of crystallinity, crystallite size is another crucial parameter that measures the physical properties of polymer composite. Throughout the XRD measurements, all composites were observed to show three main peaks on their respective XRD spectra, as shown in Figures 6 and 7. This indicates the presence of three types of crystal structures in the crystallite regions. The first peak (Peak A) was observed to appear at 2θ of 20° and it was mainly due to the presence of EVA chains in the polymer matrix [24]. The remaining two peaks, namely Peak B and Peak C, were observed to appear at 2θ of 22° and 24° respectively on the XRD spectra and it was mainly due to the presence of HDPE chains in the composite [25]. For the un-degraded composites, the sizes of the crystal structures were observed to decrease with the increase in CuO content, as shown in Table 3. This observation was consistent with the results obtained in the degree of crystallinity where the incorporation of CuO increases, the tendency for the crystal structures to be disturbed by the large CuO particles clusters will eventually be higher.

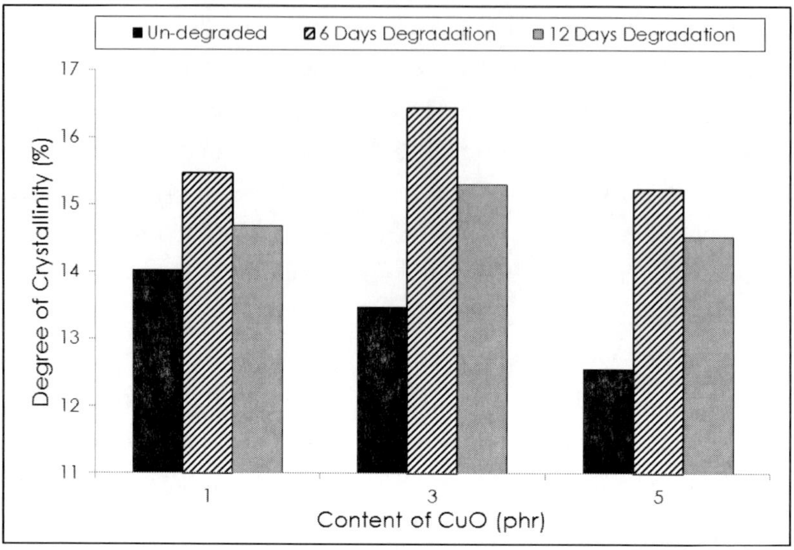

(a)

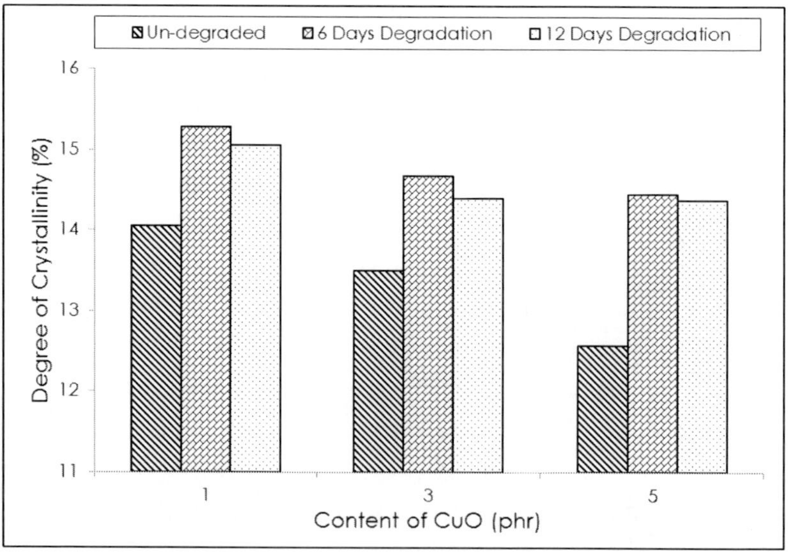

(b)

Figure 5. Degree of Crystallinity for HDPE/EVA Blends at Various CuO Compositions for (a) Ramp-120°C Degradation and (b) Ramp-100°C Degradation.

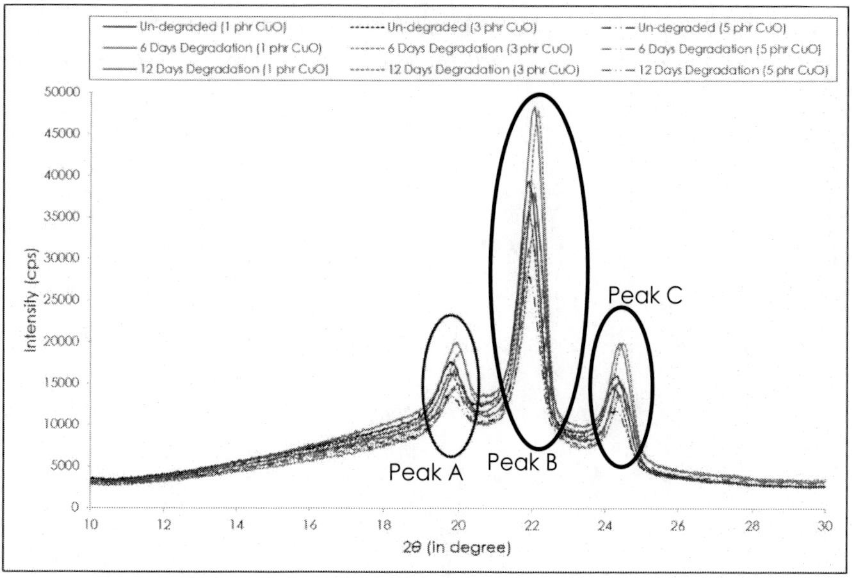

Figure 6. XRD Spectra for HDPE/EVA Blends at Various CuO Compositions for (a) Ramp-120°C Degradation.

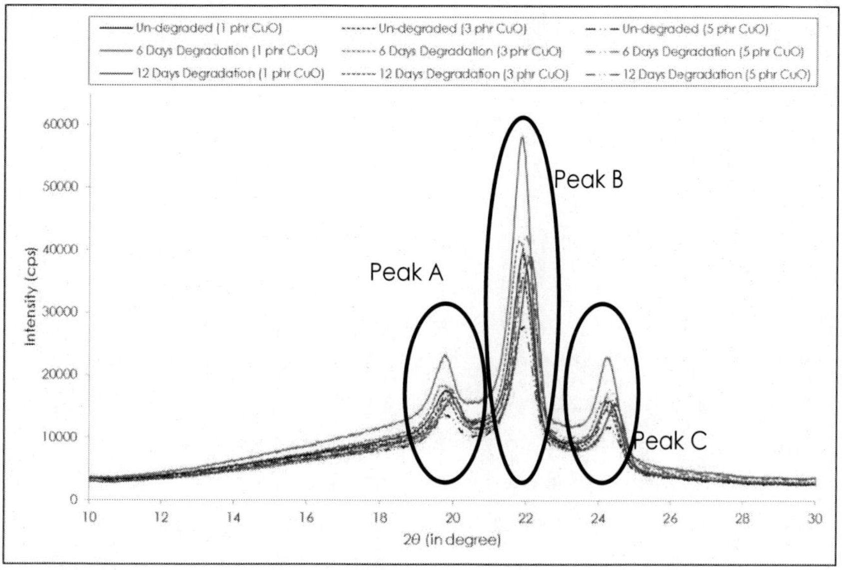

Figure 7. XRD Spectra for HDPE/EVA Blends at Various CuO Compositions for 100°C Degradation.

Table 3. Crystallites Sizes for HDPE/EVA Blends at Various CuO Compositions for Ramp-120°C and Ramp-100°C Degradation

Content of CuO (phr)	Duration of Thermal Ageing (Days)	Crystallite Sizes (Å)					
		Ramp-120°C Degradation			Ramp-100°C Ageing		
		Peak A	Peak B	Peak C	Peak A	Peak B	Peak C
1	Un-degraded	62.18	94.09	96.37	62.18	94.09	96.37
	6	63.18	95.40	107.70	75.77	112.01	110.46
	12	66.36	93.64	105.46	59.59	94.40	104.56
3	Un-degraded	60.43	90.20	99.59	60.43	90.20	99.59
	6	62.81	102.06	110.15	59.58	90.45	101.54
	12	60.48	89.65	103.62	65.36	99.43	95.94
5	Un-degraded	58.70	83.21	107.76	58.70	83.21	107.76
	6	71.90	100.69	112.34	66.42	107.91	104.11
	12	65.38	88.83	105.03	62.88	92.21	100.42

For composites that have been subjected to ramp-120°C degradation for 6 days, the sizes of the crystal structures were observed to increase. This indicates that during the course of degradation, the polymer chains in the amorphous region have dislocated from their initial positions and aligned with the crystal structures. In terms of CuO composition, the crystallite sizes were observed to increase with increment of CuO content. This could be explained through the chain scissioning of polymer chains that induced by the CuO particles. As discussed in the DSC results, certain polymer chains have been discontinued into shorter polymer chains. Since short chain polymers can easily aligned with the crystal structures as compared to that of longer polymer chains, composites with higher CuO content were expected to show larger crystallite sizes after 6 days thermal degradation. As the duration of thermal degradation was prolonged to 12 days, the crystallite sizes of the composites were observed to reduce. Furthermore, composite with 3 phr of CuO was observed to show minimum crystallite sizes after 12 days degradation. However, the composite exhibited maximum degree of crystallinity as compared to other composite. This indicates that the crystal structures in composite with 3 phr of CuO have been fractured and separated apart during the course of thermal degradation.

For composites that have been subjected to ramp-100°C degradation for 6 days, all composites possessed increase in crystallites sizes. This could also be explained by the recrystallization of polymer chains. However, composite with 3 phr of CuO was observed to exhibit minimum crystallites sizes. As the duration of degradation as prolonged to 12 days, composites with 1 and 5 phr of CuO possessed reduction in crystallites sizes. At this particular stage, composite with 3 phr was observed to show increase in crystallites sizes, indicating the enhancement in both mechanical and thermal properties of composite.

3.5. Scanning Electron Microscopy

In order to verify the impacts of ramp temperature degradation and CuO composition on the physical properties of the polymer blends, the fractured regions of the composites have been subjected to SEM analysis. Throughout the measurements, four distinct types of morphological structures were observed on the fractured surfaces of the composites. The first type of structure is known as continuity. Continuity is the smooth and continuous structure which formed due to the excellent compatibility of the blends and good intermolecular interaction between the polymer chains [26]. With this morphological structure, the composite will certainly possess good mechanical properties. The second type of structure is known as fibrils. Fibrils are the hair-liked structures which formed due to the resistance ability of the polymer matrix from elongating when the composite was subjected to straining effect [21]. The third type of structure is known as globules. The globules structures are the spherical-liked structures which formed due to the elastic effects when the polymer chains realigned themselves after the tensile elongations [21]. The forth type of structure is known as void. It is the empty "hole" which formed when the short chain polymers were further oxidized into gases (i.e., carbon dioxide) and the gasses were trapped in the polymer matrix.

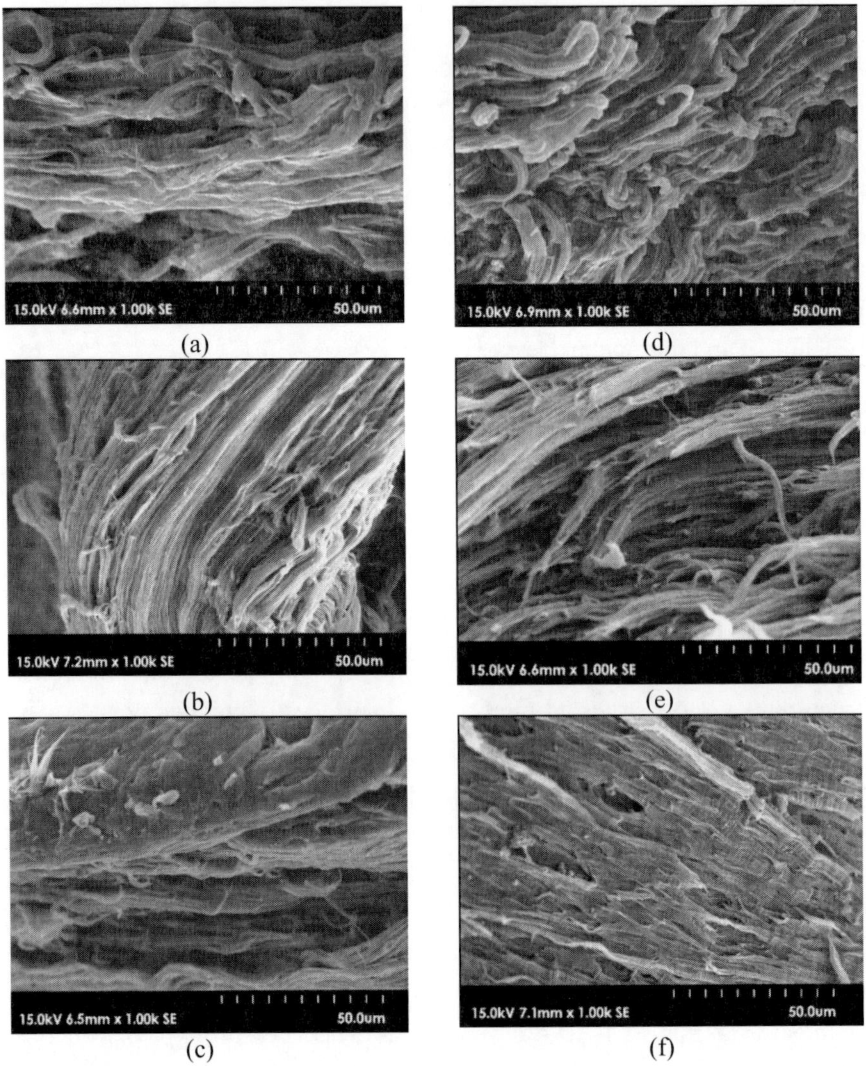

Figure 8. HDPE/EVA Blends that have been subjected to Ramp-120 °C Degradation for 6 days with (a) 1 phr CuO (b) 3 phr CuO and (c) 5 phr CuO; and for 12 days with (d) 1 phr CuO (e) 3 phr CuO and (f) 5 phr CuO.

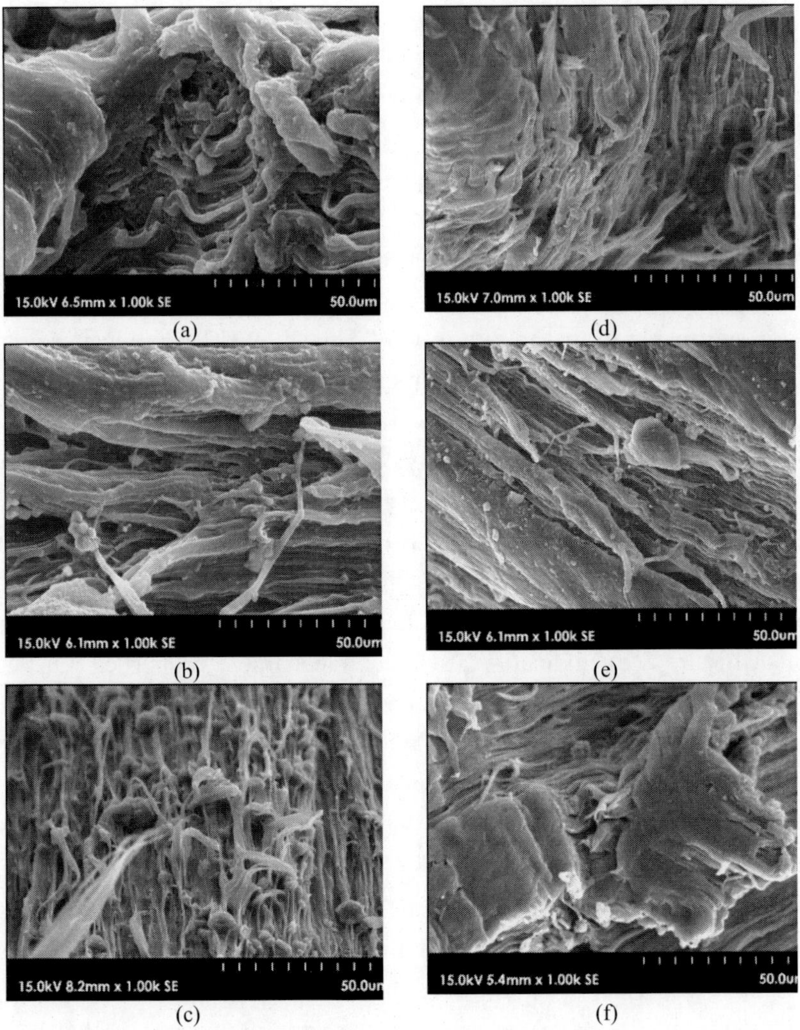

Figure 9. HDPE/EVA Blends that have been subjected to Ramp-100°C Degradation for 6 days with (a) 1 phr CuO (b) 3 phr CuO and (c) 5 phr CuO; and for 12 days with (d) 1 phr CuO (e) 3 phr CuO and (f) 5 phr CuO.

Figure 8(a), 8(b) and 8(c) show the SEM images of the composites that have been subjected to ramp-120°C degradation for 6 at various CuO compositions. By referring to Figure 8(a), it can be observed continuity was the major morphological structure that was found on the fractured surface of composite with 1 phr of CuO. This shows that the composite possessed good tensile strength during the tensile elongation. When the CuO content was increased to 3 phr, thin and long fibrils were observed on the fractured surface, as shown in Figure 8(b). This indicates that the composites possessed high elongation at break and this observation is consistent with the result obtained in the tensile test. When the CuO content was further increased to 5 phr, both continuity and fibrils were observed on the fractured surface, as shown in Figure 8(c). However, the fibrils on the fractured surface of the composite were observed to be shorter as compared to that in Figure 8(b). This shows the composites with 3 phr of CuO possessed better mechanical properties than that of 3 phr of CuO. When the duration of thermal degradation was prolonged to 12 days, composites with 1 phr of CuO showed short and thick fibrils on the fractured surface, as shown in Figure 8(d). This shows that the mechanical properties of the composite have been deteriorated due to the prolonged degradation process. For composite with 3 phr of CuO, the fibrils on the fractured surface were observed to be longer than that of 1 phr of CuO, as shown in Figure 8(e), indicating its better mechanical properties. On the other hand, it can be observed from Figure 8(f) that the voids were found on the fractured surface of composite with 5 phr of CuO. This will lead to the significant deterioration in mechanical properties.

Figures 9(a), 9(b) and 9(c) show the SEM images for composites that have been subjected to ramp-100°C degradation at various CuO compositions. For composite with 1 phr of CuO, it can be observed from Figure 9(a) that the composite showed both continuity and fibrils structures on the fractured surface. As the CuO content was increased to 3 phr, small amount of globules were observed on the fractured surface, as shown in Figure 9(b). Apart from that, the fibrils on the fractured surface were observed to be longer than that in Figure 9(a). This indicates that composite with 3 phr of CuO showed better mechanical properties that that of 1 phr. When the CuO content was further increased to 5 phr, more globules and fibrils were seen on the fractured surface, as shown in Figure 9(c). This implies that composite with 5 phr of CuO showed the higher mechanical properties as compared to the rest of the composites. When the duration of thermal degradation was prolonged to 12 days, composite with

1 phr of CuO showed more fibrils as compared to that of 6 days degradation, as shown in Figure 9(d). For composite with 3 phr of CuO (as shown in Figure 9(d)), the fibrils were observed to be longer and the amount of globules were seen to be more as compared to that of 6 days degradation. This indicates that the mechanical properties of the composite have been enhanced after the prolonged thermal degradation. Last but not least, it can be observed from Figure 9(f) that the composite with 5 phr of CuO showed think continuity and short fibrils on its fractured surface. This shows that the composite possessed relatively higher tensile strength but lower elongation at break during the tensile elongation.

CONCLUSION

This study was conducted to investigate the effects of CuO composition on the degradability of HDPE/EVA blend. Composite with 3 phr of CuO was observed to show highest mechanical properties before the thermal degradation. At lower ramp temperature, the mechanical properties were seen to increase with CuO content after the thermal degradation. When the ramp temperature was increased from 100 to 120°C, composite with 3 phr of CuO was still observed to possess better mechanical properties as compared to other composites.

In terms of melting behaviors, it was found that both the T_m and ΔH_m for the un-degraded composites increase with the content of CuO. However, maximum $T_{5\%}$ values for all the un-degraded composites took place at 3 phr of CuO content and this implies that composite with 3 phr of CuO showed highest thermal properties. After the thermal degradation, all composites showed increase in T_m and ΔH_m and the parameters were also found to increase with CuO content. On the other hand, the values of $T_{5\%}$ were found to reduce when the CuO content increased. This indicates that the thermal stabilities of the composites decreased when content of CuO increased.

For the physical properties of the composites, the degree of crystallinity and crystallites sizes for the un-degraded composites reduced in relation to the increment of CuO content. After the high temperature thermal degradation, all the composites showed increase in degree of crystallinity. Apart from that, composite with 3 phr of CuO was observed to show maximum degree of crystallinity and this also proves that the composite possesses higher melting behavior among the composites. For the surface morphologies of the composites, the continuity structures were observed to become thinner and eventually disappeared after prolonged and high temperature thermal degradation. In terms of the content of CuO, the continuity structures became thinner when the CuO content increased. However, it was observed that composite with 3 phr of CuO showed longer fibrils after high temperature degradation. This also implies that the composite showed better mechanical properties after the thermal degradation.

ACKNOWLEDGMENT

The authors would like to take this opportunity to express their sincere gratitude to Mr Foong Jee Lip from Universiti Tunku Abdul Rahman in conducting the XRD analysis.

REFERENCES

[1] Cho, M.H., Bahadur, S. *Wear* 2005, 258(5), 835.

[2] Pavličević, J., Špírková, M., Bera, O., Jovičić, M., Pilić, B., Baloš, S., Budinski-Simendić, J. *Compos. Part B-Eng.* 2014, 60, 673.

[3] Nakane, K., Kurita, T., Ogihara, T., Ogata, N. *Compos. Part B-Eng.* 2004, 35, 219.

[4] Ren, G., Hu, D., Cheng, E.W., Vargas-Reus, M.A., Reip, P., Allaker, R.P. *Int. J. Antimicrob. Agents* 2009, 33, 587.

[5] Cava, R.J. *Science* 1990, 247, 656.

[6] Hong, S.G., Wang, T.C. *J. Appl. Polym. Sci.* 1994, 52, 1339.

[7] Datta, S.K., Bhowmick, A.K., Mukunda, P.G., Chaki, T.K. *Polym. Degrad. Stab.* 1995, 50, 75.

[8] Oshima, A., Ikeda, S., Seguchi, T., Tabata, Y. *Radiat. Phys. Chem.* 1997, 49, 279.

[9] Borsova, B., Kressler, J. *Macromol. Mater. Eng.* 2003, 288, 509.

[10] Takidis, C., Bikiaris, D.N., Papageorgious, G.Z., Achilias, D.S., Sideridou, I. *J. Appl. Polym. Sci.* 2003, 90, 841.

[11] Henderson, A.M., *IEEE Elec. Insul. Magazine* 1993, 3, 30.

[12] Bee, S.T., Hassan, A., Ratnam, C.T., Tee, T.T., Sin, L.T. *Polym. Compos*. 2012, 33, 1883-1892.

[13] Selvin, T.P., Kuruvilla, J., Sabu, T. *Mater. Lett.* 2004, 58, 281.

[14] Luyt, A.S., Molefi, J.A., Krump, H. *Polym. Degrad. Stab.* 2006, 91,1629.

[15] Kim, G.M., Michler, G.H. *Polym.* 1998, 39, 5699.

[16] Osawa, Z. *Polym. Degrad. Stab.* 1988, 20, 203.

[17] Prinos, J., Bikiaris, D., Theologidis, S., Panayiotou, C. *Polym. Eng. Sci.* 1998, 38, 954.

[18] Liufu, S., Xiao, N., Li, Y. *Polym. Degrad. Stab.* 2005, 87, 103.

[19] Chien, J.C.W. *J. Phys. Chem.* 1965, 96, 4317.

[20] Kavesh, S., Schultz, J.M. *Polym. Eng. Sci.* 2004, 9, 331.

[21] Bee, S.T., Ratnam, C.T., Sin, L.T., Tee, T.T., Hui, D., Kadhum, A.A.H., Rahmat, A.R., Lau, J. *Compos. Part B-Eng.* 2014, 63, 141.

[22] Johan, M.R., Leo, B.F. *Ionics* 2010, 16, 335.

[23] Tee, T.T., Sin, L.T., Gobinath, R., Bee, S.T., Hui, D., Rahmat, A.R., Kong, I., Fang, Q. *Compos. Part B-Eng.* 2013, 47, 238.

[24] Pang, H., Zhong, G., Wang, Y., Xu, J., Li, Z., Lei, J., Chen, C., Ji, X. *J. Polym. Res.* 2012, 19, 1.

[25] Chattopadhyay S, Chaki TK and Bhowmick AK. Structural characterization of electron beam crosslinked thermoplastic elastomeric films from blends of polyethylene and ethylene-vinyl acetate copolymer. *J Appl Polym Sci.* 2001, 81(8): 1936-1950.

[26] Ng H.M., Bee S.T., Ratnam C.T., Sin L.T., Phang Y.Y., Tee T.T., Rahmat A.R.. Effectiveness of trimethylopropane trimethacrylate for the electron-beam-irradiation-induced cross-linking of polylactic acid. *Nucl Instrum Meth. B* 2014, 319: 62-70.

INDEX

D

E

F

G

H

I

L

M

N

O

P

R

V

W

X

Y

Z